Novel Three-state Quantum Dot Gate Field Effect Transistor

Supriya Karmakar

Novel Three-state Quantum Dot Gate Field Effect Transistor

Fabrication, Modeling and Applications

Supriya Karmakar
Intel Corporation
Hillsboro, Oregon
USA

ISBN 978-81-322-3490-6 ISBN 978-81-322-1635-3 (eBook)
DOI 10.1007/978-81-322-1635-3
Springer New Delhi Heidelberg New York Dordrecht London

Softcover reprint of the hardcover 1st edition 2014

Printed on acid-free paper

Springer is part of Springer Science+Business Media (www.springer.com)

To my parents
Manik Chandra Karmakar
and
Santwana Karmakar

Preface

This book presents the fabrication and circuit modeling of a quantum dot gate field-effect transistor (QDGFET) and a quantum dot gate NMOS inverter (QDNINV). A conventional metal-oxide-semiconductor field-effect transistor (MOSFET) conducts when the applied gate voltage is more than the threshold voltage of the device. So a MOSFET acts as a switch which cannot conduct below its threshold voltage but conducts beyond its threshold voltage. A QDGFET produces three states in its transfer characteristics: off, on, and a low-current saturation state, known as the intermediate state ("i"), because of the presence of quantum dots in the gate region. A self-consistent solution of Schrödinger and Poisson equations can explain the generation of the intermediate state between off and on states of the QDGFET.

The long-channel QDGFETs were fabricated on a p-type (100) silicon wafer as well as on a silicon-on-insulator wafer. Two different types of quantum dots (SiOx-cladded Si and Germanium-cladded Ge) are site-specifically self-assembled on top of the 20 Å silicon dioxide gate insulator grown by thermal oxidation and II-VI ZnS-ZnMgS gate insulator grown by the metal organic chemical vapor deposition (MOCVD) technique. In QDNINV, SiOx-cladded Si dots are self-assembled on top of thermally grown silicon dioxide in the gate region of the QDGFETs in the inverter circuit.

This book also introduces the development of a circuit model of QDGFET based on the Berkley Short Channel IGFET model (BSIM). Different ternary logic circuits based on QDGFET are also investigated in this book. Advanced circuits like three-bit and six-bit analog-to-digital converter (ADC) and digital-to-analog converter (DAC) are also simulated.

Acknowledgments

I wish to express my first and foremost gratitude to my major advisor Dr. Faquir Jain without whom this task was inconceivable. This work was supported by the Office of Naval Research Contracts and an NSF grant. I would like to gratefully acknowledge the help of my associate advisor committee, Dr. John Chandy, Dr. Rajeev Bansal, Dr. John Ayers, and Dr. Lei Wang, for their valuable advices and comments.

As with any collaborative work, I am also indebted to other laboratories such as the Microsystems Technology Laboratories (MTL) at the Massachusetts Institute of Technology (MIT), Cornell Nanoscale Science & Technology Facility (CNF) at Cornell University, Center for Nanoscale Systems (CNS) at Harvard University, and Yale University.

I owe many thanks to all lab members of the Micro/Optoelectronic Laboratory, Department of Electrical and Computer Engineering, University of Connecticut. I am grateful to Dr. Evan Heller for his valuable advices on device modeling. I would like to thank all faculty members of the Electrical and Computer Engineering Department as well as department secretaries from whom I was privileged to learn. I would also like to extend thanks and appreciations to my colleagues at Intel Corporation.

Words are not enough to express my deep sense of gratitude to my parents for their support. Without their sacrifices, I would have never come this far. Also, I would like to extend my heartfelt thanks to my sister and my brother-in-law for their support and guidance during my study.

Finally, I would like to thank my wife, who walked beside me, supported me along the way constantly, and encouraged me to write a book. I thank her for her patience, understanding, and love.

Contents

Chapter 1
Introduction: Multistate Devices and Logic

This chapter discusses the multi-valued logic and different negative tunneling devices to implement this multi-valued logic. Problems of different negative resistance devices and possible solution using quantum dot gate FET (QDGFET) are studied. Advantages of QDGFET based circuits over the conventional FET based circuits are also explored.

A metal-oxide-semiconductor field-effect transistor (MOSFET) produces two states based on its applied gate voltage. When the gate voltage is below the threshold voltage, the transistor is off, and when the gate voltage is more than the threshold voltage, an inversion channel forms below the gate and current conduction occurs between the source and the drain terminals of the device. This allows MOSFET to be suitable only for binary logic.

The trend predicted by Moore's law has called for these MOS devices to shrink in half every 18–24 months [1, 2]. As feature sizes have started to approach sub-22-nm regime, several issues have begun to make further miniaturization difficult [3–5]. As transistors have decreased in size, the thickness of the gate dielectric needs to decrease to increase the gate capacitance and thereby the drive current and device performance [6–9]. As the gate dielectric thickness scales down below 2 nm, leakage currents due to the tunneling of charge carriers increase drastically, leading to unwieldy power consumption and reduced device reliability [10–12]. Another approach to increase integration is to increase the bit handling capacity of a fabricated device. Traditional MOSFETs can only process one bit at a time.

Higher level of bit density can be achieved using multivalued logic (MVL) [13]. Multivalued logic is defined as a nonbinary logic and involves the switching between more than two states.

Research is ongoing towards the development of devices that are suitable for multivalued operation. Resonant tunneling transistors (RTTs) [14–16] and resonant tunneling diodes (RTDs) [17, 18] are the major promising semiconductor devices for MVL applications. All these devices produce negative differential resistance (NDR) in the specific voltage range of operations.

S. Karmakar, *Novel Three-state Quantum Dot Gate Field Effect Transistor: Fabrication, Modeling and Applications*, DOI 10.1007/978-81-322-1635-3_1,

1.1 Resonant Tunneling Diode (RTD)

A resonant tunneling device is similar in construction to a p-n diode with two terminals [19]. The device has a quantum well layer separated from two conducting regions by quantum barriers (on the order of 50–100 Å thick). The barriers can be AlAs, for example, while the well and conduction regions can be InGaAs.

Electrons tunnel through two barriers separated by a well when flowing from the source to the drain in a resonant tunneling diode, which is also known as quantum mechanical tunneling [17]. The energy level in the well is quantized because the well is in the order of de Broglie wavelength. The flow of electrons is controlled by diode bias. This matches the energy levels of the electrons in the source to the quantized level in the well so that electrons can tunnel through the barriers. When the energy levels are equal, a resonance occurs, allowing electron flow through the barriers as shown in Fig. 1.1b. No bias or too much bias, in Fig. 1.1a, c, respectively, yields an energy mismatch between the source and the well and thus no conduction.

As bias is increased from zero across the RTD, the current increases, reaching a maximum before it decreases. This corresponds to the off, on, and off states. Figure 1.2 shows the I-V characteristic of a resonant tunneling diode. This makes simplification of conventional transistor circuits possible by substituting a pair of RTDs for two transistors. For example, two back-to-back RTDs and a transistor form a memory cell, using fewer components, less area, and power compared to a conventional circuit. The potential application of RTDs is to reduce the component count, area, and power dissipation of conventional transistor circuits by replacing some, though not all, transistors.

RTDs have allowed us to realize certain applications that will be beyond the capability of CMOS technology [20, 21]. These low-power, high-speed, and small devices are especially important as we continue to scale down to the size of atoms, where heat and parasitic effects are major problems. However, in order for RTDs to reach its full potential, more mature fabrication techniques are needed. Precise barrier thickness control is needed to ensure uniformity across the whole wafer. Also, the output power of RTDs is limited. More research is needed to help realize RTD circuits without an amplifier or other drivers. This will minimize the power and area of the integrated circuit (IC).

1.2 Resonant Tunneling Transistor (RTT)

A resonant tunneling transistor [22] is similar in construction to the resonant tunneling device except, in addition to a source and a drain, that it has a gate electrode attached to the well region [23]. Electrons are injected into the device through the source electrode. An applied gate voltage adjusts the energy of the well levels with respect to the energy range of the occupied conduction states in the source. Carriers tunnel through the barrier into the well from the source when the energy of a well state falls within the energy range of the occupied conduction

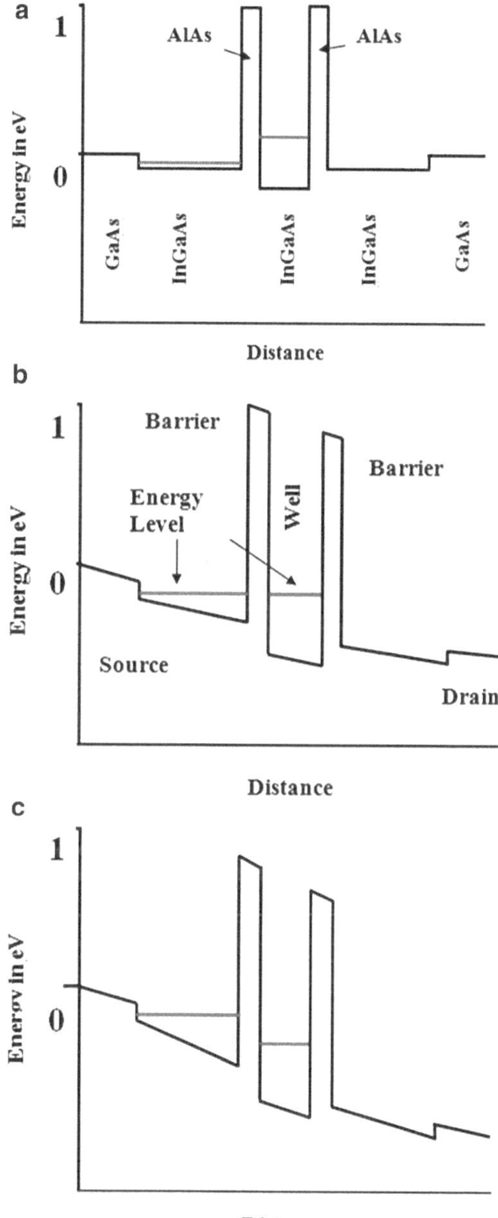

Fig. 1.1 Resonant tunneling diode (RTD): (**a**) with no applied bias, the source and well energy levels are not matched; no conduction. (**b**) Small bias causes matched energy levels (resonance); conduction results. (**c**) Further bias mismatches energy levels, decreasing conduction

states in the source [24, 25]. With a voltage applied from the source to drain, the charge then flows into the drain.

All the devices that have been proposed so far have more serious problems than complementary metal-oxide-semiconductor (CMOS) devices, such as excess leakage current, being unable to operate at room temperature, or requiring complex fabrication

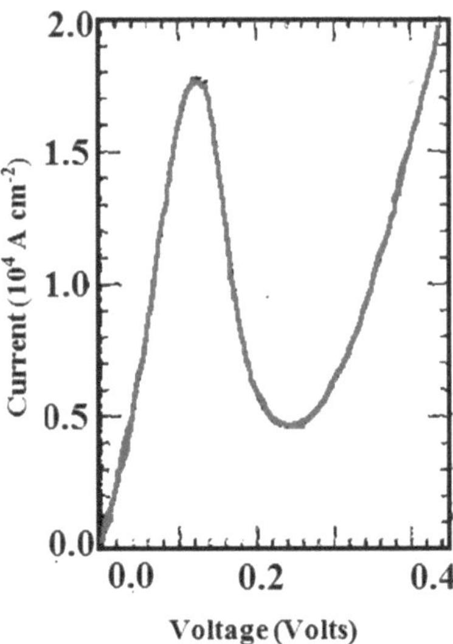

Fig. 1.2 RTD I-V characteristic

processes. These devices are not matured enough for practical MVL circuit implementation. Research is ongoing to find a better combination of stack materials for resonant tunneling devices. A common problem in the NDR devices is always how to reduce the valley current, which is due to the unintentional excess leakage current.

1.3 Quantum Dot Gate Field-Effect Transistor (QDGFET)

In a quantum dot gate field-effect transistor (QDGFET) [26], resonant tunneling occurs in the gate insulator region which consists of quantum dots. This results in the manifestation of an intermediate state which is more stable in a QDGFET because of the reduced charge leakage through the cladding of the QDs in the gate region. Thus, QDGFETs provide the capability to process three states in a single device and provide higher levels of overall bit density in a circuit for a given feature size. Besides this QDGFET can be fabricated using conventional CMOS process technology. The integration problem for other devices with existing CMOS technology will be solved in QDGFETs.

This book discusses the fabrication, modeling, and characterization of the three-state quantum dot gate field-effect transistors (QDGFETs). A device model based on self-consistent solution of Schrödinger and Poisson equations explained the three-state behavior of QDGFET. This thesis also introduced a circuit model based on Berkeley short-channel IGFET model (BSIM) for integrated circuit design. The extension of QDGFET circuit model for sub-25-nm range is also described in this work. Implementation of different ternary logics and advanced circuits like compact comparators, three-bit and six-bit analog-to-digital converters (ADCs) and digital-to-analog converters (DACs) based on QDGFET is also demonstrated in this book.

References

1. Moore, G.E.: Cramming more components onto integrated circuits. Electronics **38**(8), (1965)
2. Moore, G.E.: No exponential is forever: but 'forever' can be delayed! In: Solid-State Circuits Conference, 2003. Digest of Technical Papers. ISSCC. 2003 I.E. International, University of Pennsylvania, vol. 1, pp. 20–23 (2003)
3. Nowak, E.J.: Maintaining the benefits of CMOS scaling when scaling bogs down, pp. 169–180. J. Res. Dev., 46 (2002)
4. Theis, T.N.: Beyond the silicon transistor: personal observations. Comput. Sci. Eng. **5**, 25–29 (2003)
5. Borkar, S.: Design perspectives on 22 nm CMOS and beyond. In: Design Automation Conference, San Francisco, USA, pp. 93–94, 26–31 July 2009
6. Goto, M., Kawanaka, S., Inumiya, S., Kusunoki, N., Saitoh, M., Tatsumura, K., Kinoshita, A., Inaba, S., Toyoshima, Y.: The study of mobility-tin, trade-off in deeply scaled high-k/metal gate devices and scaling design guideline for 22 nm-node generation. In: VLSI Technology, 2009 Symposium on, Honolulu, HI, pp. 214–215, 16–18 June 2009
7. Ru Huang, Han Ming Wu, Jin Feng Kang, De Yuan Xiao, Xue Long Shi, Xia an, Yu Tian, Run Sheng Wang, Liang Liang Zhang, Xing Zhang, et al.: Challenges of 22 nm and beyond CMOS technology. Sci. China Ser. F Inf. Sci. **52**(9), 1491–1533
8. Thompson, S., et al.: A 90 nm logic technology featuring 50 nm strained silicon channel transistors, 7 layers of Cu interconnects, low k ILD, and 1 μm^2 SRAM cell. In: IEDM Technical Digest, pp. 61–64, Dec 2002
9. Mistry, K., et al.: A 45nm logic technology with high-k+metal gate transistors, strained silicon, 9 Cu interconnect layers, 193nm dry patterning, and 100% Pb-free packaging. In: IEDM Technical Digest, pp. 247–250, Dec 2007
10. Bai, P., et al.: A 65nm logic technology featuring 35nm gate lengths, enhanced channel strain, 8 Cu interconnect layers, low-k ILD and 0.57 μm^2 SRAM cell. In: IEDM Technical Digest, pp. 657–660, Dec 2004
11. Lim, H.K., Fossum, J.G.: Threshold voltage of thin-film Silicon-on-insulator (SOI) MOSFET's. IEEE Trans. Electron Devices **30**(10), 1244–1251 (1983)
12. Colinge, J.P.: Transconductance of silicon-on-insulator MOSFETs. IEEE Electron Device Lett. **EDL-6**, 573–574 (1985)
13. Micheel, L.J., Taddiken, A.H., Seabaugh, A.C.: Multiple-valued logic computation circuits using micro- and nanoelectronics devices. In: Proceedings of 23rd IEEE International Symposium on Multiple- Valued Logic, Sacramento, California, USA, pp. 164–169 (1993)
14. Seabaugh, A.C., Frensley, W.R., Randall, J.N., Reed, M.A., Farrington, D.L., Matyi, R.J.: Pseudomorphic bipolar quantum resonant-tunneling transistor. IEEE Trans. Electron Devices, **36**(10), 2228–2234 (1989)
15. Stock, J., Malindretos, J., Indlekofer, K.M., Pottgens, M., Forster, A., Luth, H.: A vertical resonant tunneling transistor for application in digital logic circuits. IEEE Trans. Electron Devices, **48**(6), 1028–1032 (2001)
16. Capasso, F., Kiehl, R.A.: Resonant tunneling transistor with quantum well base and high – energy injection: a new negative differential resistance device. J. Appl. Phys. **58**(3), 1366–1368 (1985)
17. Lin, H.C.: Resonant tunneling diodes for multi-valued digital applications. In: Proceedings of 24th IEEE International Symposium on Multiple –Valued Logic, Boston, Massachusetts, USA, pp. 188–195 (1994)
18. Forster, A.: Resonant tunneling diodes: the effect of structural properties on their performance. Adv. Solid State Phys. **33**, 37–62 (1993)
19. Waho, T., Chen, K.J., Yamamoto, M.: Resonant-tunneling diode and HEMT logic circuits with multiple thresholds and multi-level output. IEEE J. Solid-State Circuits **33**(2), 268–274 (1998)
20. Mazumder, P., Kulkarni, S., Bhattacharya, M., Sun, J.P., Haddad, G.I.: Digital circuit applications of resonant tunneling diodes. Proc. IEEE **86**(4), 664–686 (1998)

21. van der Wagt, J.P.A., Tang, H., Broekaert, T.P.E., Seabaugh, A.C., Kao, Y.-C.: Multibit resonant tunneling diode SRAM cell based on slew-rate addressing. IEEE Trans. Electron Devices **46**(1), 55–62 (1999)
22. Waho, T.: Resonant tunneling transistor and its application to multiple-valued logic circuits. In: Proceedings of 25th IEEE International Symposium on Multiple-Valued Logic, Bloomington, Indiana, USA, pp. 130–138 (1995)
23. Chen, W.L., Mums, G.0., Davis, L., Bhattacharya, P.K., Haddad, G.I.: The growth of resonant tunneling hot electron transistors using chemical beam epitaxy. In: The 4th International Conference in Chemical Beam Epitaxy, section S-7, Nara, Japan, July 1993
24. Futatsugi, T., Yamaguchi, Y., Ishii, K., Imamura, K., Muto, S., Yokoyama, N., Shibatomi, A.: A resonant tunneling bipolar transistor (RBT): a proposal and demonstration for new functional device with high current gains. In: Technical Digest IEDM, p. 286, Dec 1986
25. Seabaugh, A.C., Frensley, W. R., Kao, Y.C., Randall, J.N., Reed, M.A.: Quantum-well resonant tunneling transistors. In: The Proceedings of the 1989 I.E. Come11 Conference, Ithaca, p. 255
26. Jain, F.C., Heller, E., Karmakar, S., Chandy, J.: Device and circuit modeling using novel 3-state quantum dot gate FETs. In: International Semiconductor Device Research Symposium, College Park, 12–15 Dec 2007

Chapter 2
Quantum Dot Gate Field-Effect Transistor: Device Structures

This chapter introduces quantum dot gate field-effect transistors (QDGFETs). Different types of quantum dots (SiO_x-cladded Si and GeO_x-cladded Ge) are self-assembled on different kinds of gate insulator like silicon dioxide as well as high-κ dielectric. Different kinds of substrates (silicon and silicon-on-insulator) are also introduced as different material systems.

2.1 Metal-Oxide-Semiconductor Field-Effect Transistors (MOSFETs)

A metal-oxide-semiconductor field-effect transistor (MOSFET) is a four-terminal electronic device in which the conductivity of the channel between the source and the drain terminals is controlled by the voltage applied to the gate terminal [1–3]. On a MOSFET, the metallic or polysilicon gate is isolated from the channel by a thin layer of silicon dioxide (Fig. 2.1) [4]. Although the bottom of the insulating layer is in contact with the p-type silicon substrate, the physical processes which occur at this interface dictate that free electrons will accumulate in the interface, inverting the p-type material and spontaneously forming an n-type channel [5, 6]. Thus, a conduction path exists between the diffused n-type source and drain regions. Figure 2.2 shows the transfer characteristic (I_D–V_{GS}) of a conventional MOSFET, and Fig. 2.3 shows the output characteristics (I_D–V_{DS}) of a conventional MOSFET.

2.2 Scaling Issues

Different parameters of MOSFETs such as channel length, channel width, gate oxide thickness, and other dimensions have been scaled down for improving performance and density. When the feature sizes approach sub-22-nm range,

S. Karmakar, *Novel Three-state Quantum Dot Gate Field Effect Transistor: Fabrication, Modeling and Applications*, DOI 10.1007/978-81-322-1635-3_2,

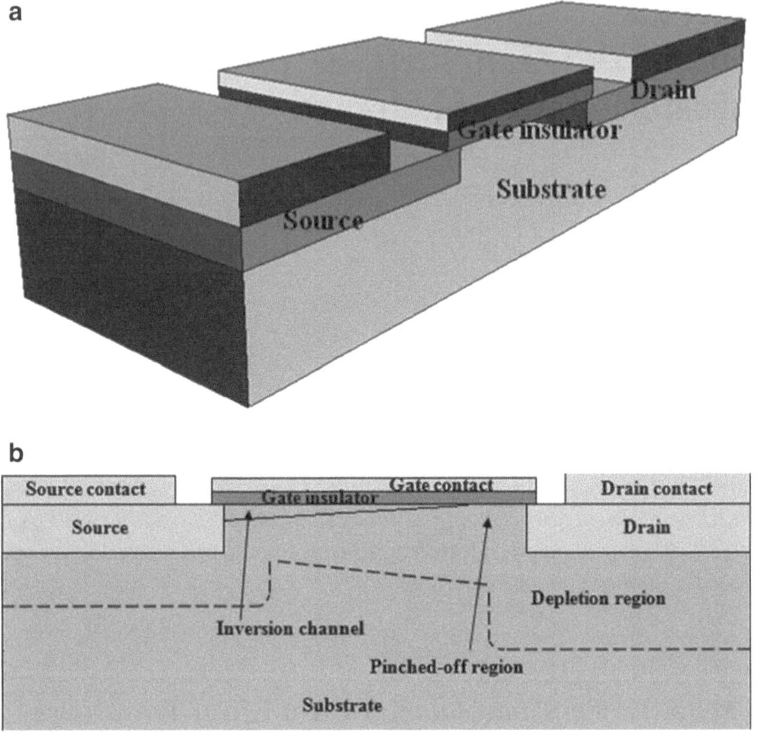

Fig. 2.1 (**a**) Device structure and (**b**) cross-sectional schematic of a conventional metal-oxide-semiconductor field-effect transistor (MOSFET)

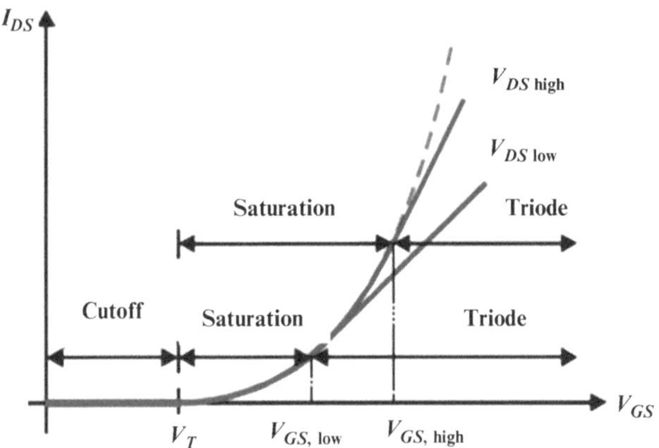

Fig. 2.2 Transfer characteristic of a metal-oxide-semiconductor field-effect transistor (MOSFET)

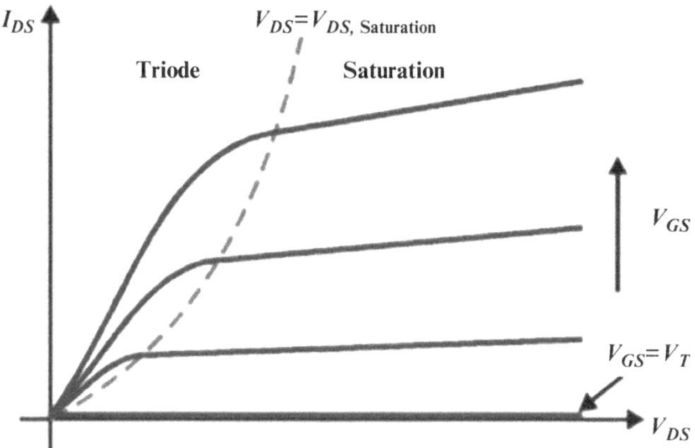

Fig. 2.3 Output characteristic of a metal-oxide-semiconductor field-effect transistor (MOSFET)

several issues arise to make further scaling difficult. As transistors have decreased in size, the thickness of the gate dielectric needs to decrease to increase the gate capacitance and thereby the drive current and the device performance.

As the gate dielectric thickness scales down below 2 nm, leakage currents due to the tunneling of charge carriers increase drastically, leading to unwieldy power consumption and reduced device reliability [7–10]. Besides the gate dielectric thickness, there are many other scaling issues, such as sensitivity to doping fluctuations, interface state and surface charges, different kinds of short-channel effects, quantum confinement in the inversion layer, and source-drain series resistance (because of thinner junction depth and others) which can affect transistor characteristics in the sub-nm range. With planar technology, a channel length of 20 nm is feasible, but for practical application, most likely a 10 nm channel length is the scaling limit, even for 3-dimensional structures.

Different parameters like device structures [11–13], gate dielectric materials [14–16], substrate doping, and source-drain doping profile have been proposed to control the short-channel effects and to improve MOSFET performance.

Development efforts have focused on finding a material [17, 18] with a requisite high dielectric constant that allows increased gate capacitance without the concomitant leakage effects and can be easily integrated into the existing Si manufacturing process [19]. Other key considerations include energy band engineering of silicon (which may alter leakage current), film morphology, thermal stability, maintenance of high mobility of charge carriers in the channel, and minimization of electrical defects in the film/interface. It is expected that defect states in the high-κ dielectric can influence its electrical properties [20, 21].

Another major short-channel effect is punch through [22, 23], which is caused by the touching of depletion region in the source and drain junction. When the channel length decreases, the drain depletion region and source depletion region touch each other, and a low-field region is generated through the substrate which is

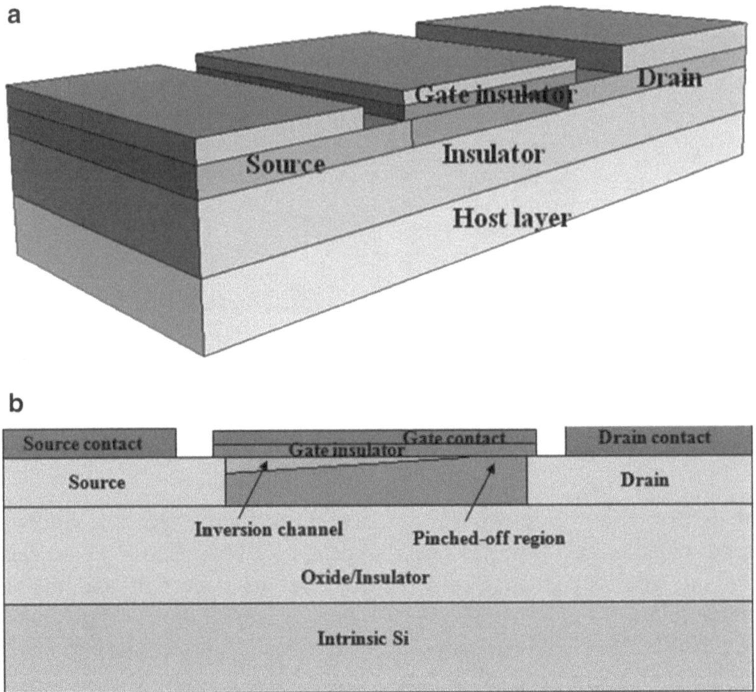

Fig. 2.4 (**a**) Device structure and (**b**) cross-sectional schematic of a conventional metal-oxide-semiconductor field-effect transistor (MOSFET) on silicon-on-insulator (SOI) substrate

known as the saddle point. This regime of operation is known as punch through. Under punch-through condition, charge carriers from source to drain transfer through the saddle point, and the drain-to-source current is controlled by the drain-to-source voltage (V_{DS}) instead of gate-to-source voltage (V_{GS}). The punch-through problem can be avoided by increasing the substrate doping of the MOSFET; however, this will increase the threshold voltage of the device as well as decrease the carrier mobility inside the channel of the FET. On the other hand, thin body (partially depleted or fully depleted) can be used to reduce the punch-through effect such that the channel can be doped lightly. Silicon-on-insulator (SOI) substrate can be advantageous [24, 25] for improved MOSFET scaling due to its thin body. SOI wafers have a top silicon layer with different insulator materials and holding substrates. The top silicon layer of an SOI wafer is a high-quality single-crystal material that is suitable for high-performance and high-density integrated circuits.

Figure 2.4 shows the cross-sectional schematic of a MOSFET on SOI wafer. The thin body of SOI wafer can alleviate most problems of punch through such that the channel can be doped lightly. The subthreshold swing which is because of diffusion charge carriers between source and drain also decreases. Other advantages include good isolation due to the buried oxide layer which reduces the substrate capacitance and increase the speed of the device. Device isolation is also much easier which can improve the circuit density. Device isolation also eliminates the latch-up problem in CMOS circuits.

2.3 Quantum Dot

A quantum dot represents a potential well that confines electrons in three dimensions to a region of the order of the electron's de Broglie wavelength in size, a few nanometers in a semiconductor. All isolated atoms are quantum dots, but multimolecular combinations can have this characteristic as well. Because of the confinement, electrons in the quantum dot have quantized, discrete energy levels, much like an atom. For this reason, quantum dots are sometimes called "artificial atoms." The energy levels can be controlled by changing the size and shape of the quantum dot and the depth of the potential. Quantum dots can be fabricated using mainly three different techniques such as lithography [26, 27], colloidal synthesis [28, 29], and epitaxial method such as MOCVD [30, 31] and MBE [32, 33]. With lithography, small features are fabricated using an electron or ion beam method. In colloidal synthesis, quantum dots are precipitated on the substrate from quantum dot solution by chemical reactions. In epitaxial growth, quantum dots are formed because of lattice mismatch between substrate and the crystallizing material.

2.4 Quantum Dot Gate Field-Effect Transistor (QDGFET)

2.4.1 Device Structure

2.4.1.1 SiO_x-Cladded Si Quantum Dots on Top of SiO_2 Tunnel Gate Insulator

FETs with quantum dot gates have been reported in the context of floating gate nonvolatile memories by a number of investigators following the work of Tiwari et al. [34]. The structural difference between a floating gate memory and a quantum dot gate three-state FET (QDGFET) [35] is the absence of a control gate insulator in the quantum dot gate FETs. No charge can be stored in the gate region because of the absence of a control gate insulator. As a result, the threshold voltage of a QDGFET varies with the applied gate voltage which is absent in floating gate memories. In addition, quantum dot gate FETs, consisting of two layers of quantum dots, show three stable states ("0," "i," and "1") in their transfer characteristics, whereas quantum dot floating gate memories show the threshold voltage shift in their transfer characteristics only after a write operation.

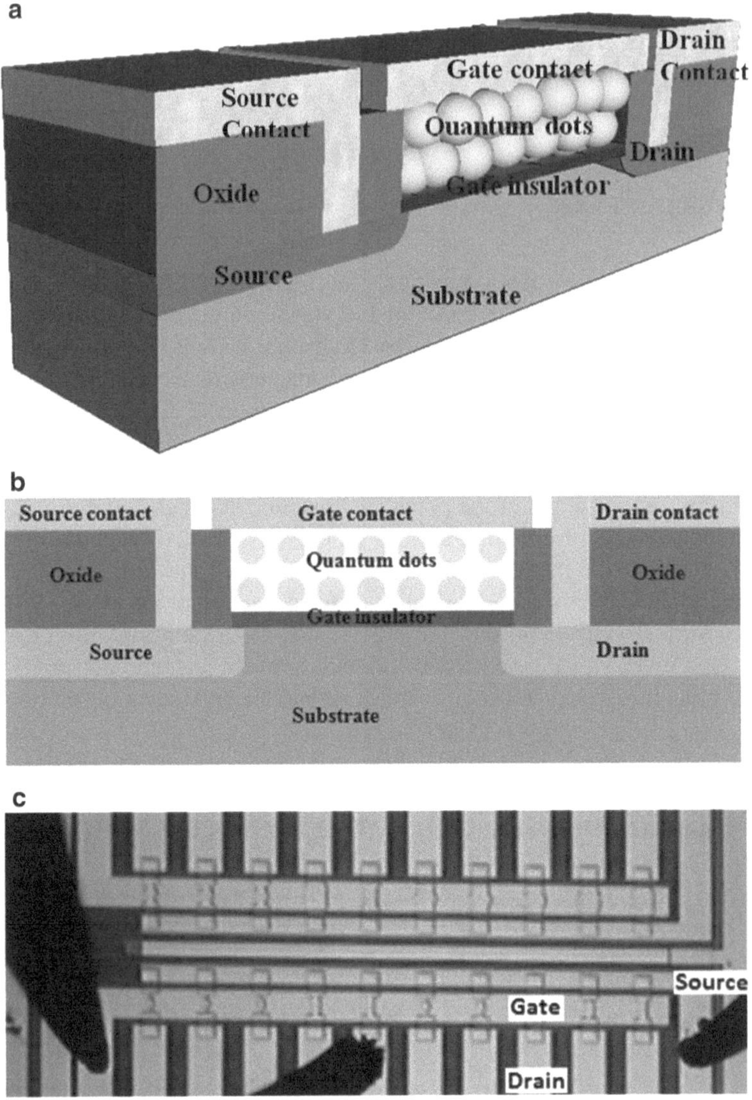

Fig. 2.5 (**a**) Device structure, (**b**) cross-sectional schematic, and (**c**) *top view* of a quantum dot gate field-effect transistor (QDGFET)

Quantum dots are created by using a site-specific self-assembly process [36–38]. In a QDGFET, quantum dots are assembled in the gate region (Fig. 2.5). Figure 2.6 shows the transfer characteristics of a fabricated QDGFET where SiO_x-cladded Si quantum dots are self-assembled on top of the 20-Å SiO_x gate insulator. Because of the presence of quantum dots in the gate region, QDGFETs produce an intermediate state between the two stable states. The existence of this intermediate state in the QDGFET makes it useful to handle more bits at a time in multivalued logic circuits.

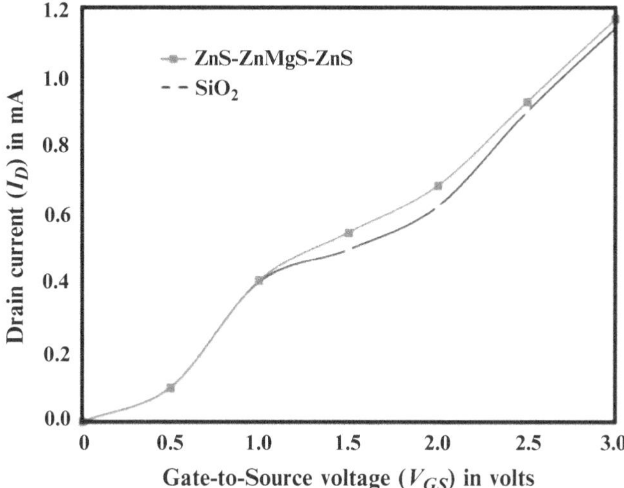

Fig. 2.6 Transfer characteristic of a quantum dot gate field-effect transistor (QDGFET)

Three-state behavior has been observed in the following structures:

(a) SiO_x-cladded Si dots on top of SiO_2 gate insulator (Fig. 2.5)
(b) GeO_x-cladded Ge dots on top of ZnS-ZnMgS gate insulator (Fig. 2.8)
(c) SiO_x-cladded Si dots on top of SiO_2 gate insulator in SOI substrate (Fig. 2.10)
(d) Thin layer of SiN on top of SiO_x-cladded Si dots on top of SiO_2 gate insulator (Fig. 2.11)

In this chapter, we demonstrate a QDGFET using different types of quantum dots (SiO_x-cladded Si/GeO_x-cladded Ge) on top of silicon dioxide, as well as on a high-k dielectric in the gate region. Two types of substrates (*p*-silicon <100 > and silicon-on-insulator) have been used. The punch-through problem for short-channel devices can be solved using SOI wafers. The device isolation feature and subthreshold swing can also be improved using SOI substrates (Fig. 2.7).

The key feature of QDGFETs that allow multi-bit processing is that they produce one intermediate state between the normal two stable on and off states. In this intermediate state, the drain current (I_D) does not vary or varies slowly with the gate-to-source voltage (V_{GS}) and produces a flat region in the transfer characteristic of the QDGFET.

Design of very compact comparator circuits based on these three-state QDGFETs is also possible, and it can reduce the comparator size by a factor of 16 compared to traditional Si MOSFET-based comparators. Analog-to-digital converters (ADCs) and digital-to-analog converters (DACs) based on this comparator circuit are also simulated. The presence of an intermediate state between two stable states of QDGFETs can make them promising circuit elements in the semiconductor industry.

The presence of the silicon-on-insulator substrate does not affect the three-state behavior of the QDGFET and presents a very distinct intermediate state “i” in the transfer characteristics of a silicon FET having SiO_X-Si quantum dots on the gate

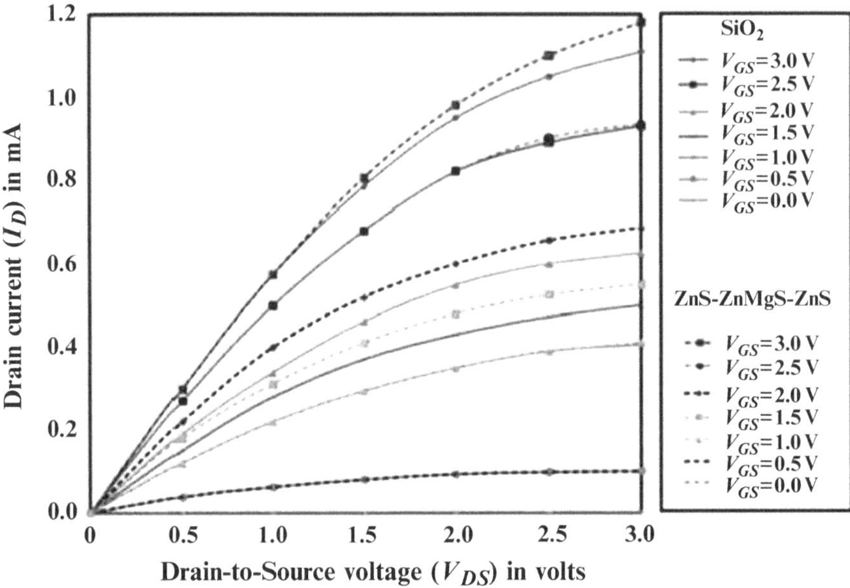

Fig. 2.7 Output characteristic of a quantum dot gate field effect transistor (QDGFET) having SiO_2 and ZnS-ZnMgS-ZnS as gate insulator

insulator. The improved on-to-off ratio will increase the device speed as well as the device performance. The noise margin between different states also increases.

2.4.1.2 GeO_x-Cladded Ge Quantum Dots on Top of High-κ Gate Dielectric

Figure 2.8 shows the schematic cross section of a QDGFET having self-assembled GeO_x-cladded Ge dots in the gate region on the p-Si substrate between the source and drain. Two layers of GeO_X-cladded Ge quantum dots (QDs) are self-assembled on the lattice-matched multi-stack gate insulator layer.

The use of lattice-matched II–VI gate dielectric reduces the interface states at the channel-gate insulator boundary. In this work, a multilayer stack of lattice-matched ZnS-ZnMgS is used as the gate insulator in this QDGFET. Figure 2.9 shows the bandgap energy versus lattice constant for different materials. The lattice constant for Si (5.43 Å at 300 K) and the lattice constant of ZnS (5.42 Å at 300 K) are almost same. We use $Zn_{1-x}Mg_xS$ as a high bandgap (for $x = 0.05$, e.g., ~3.8 eV for $Zn_{1-x}Mg_xS$ with respect to silicon which is 1.12 eV at 300 K) and high-κ (ZnS zinc blende, 8.9; wurtzite, 9.6; SiO_2, 3.9) dielectric as a gate insulator. Since ZnS has almost same lattice constant as silicon (lattice constant of Si is 5.43 Å at 300 K and that of ZnS is 5.42 Å at 300 K), it is used as a buffer layer between Si and $Zn_{1-x}Mg_xS$ gate insulator to reduce the interface states at the channel-gate insulator boundary.

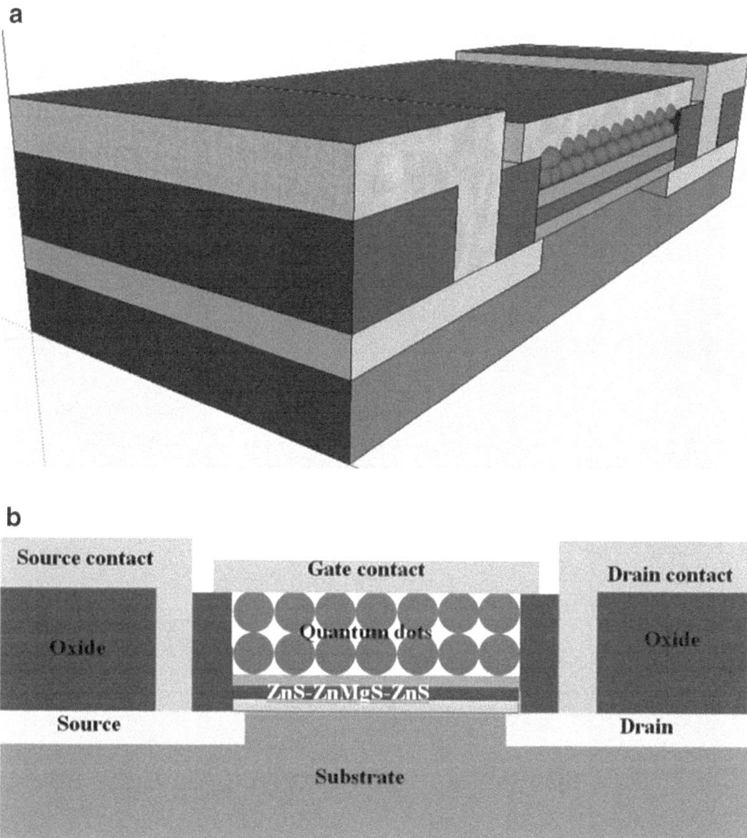

Fig. 2.8 (**a**) Device structure and (**b**) cross section of a QDGFET having self-assembled GeO_x-cladded Ge dots on top of lattice-matched high-κ gate insulator on the p-Si substrate between source and drain

The lattice-matched gate dielectric reduces the interface states at the channel-gate insulator boundary, and the high-κ of this insulator makes it a good gate dielectric.

The lattice-matched gate insulator does not affect the three-state behavior of the QDGFET and generates a very distinct intermediate state "i" in the transfer characteristics of a Si FET having GeO_X-Ge quantum dots on the lattice-matched gate insulator.

2.4.1.3 SiO_x-Cladded Si Quantum Dots on Top of SiO_2 in SOI Substrate

Figure 2.10 shows the cross-sectional schematic of a QDGFET on SOI substrate having self-assembled SiO_x-cladded Si dots in the gate region on the p-type silicon between the source and the drain region. The quantum dot layers are self-assembled on top of the gate insulator. Aluminum is used as the gate contact, and gold arsenic is used as the source-drain contacts.

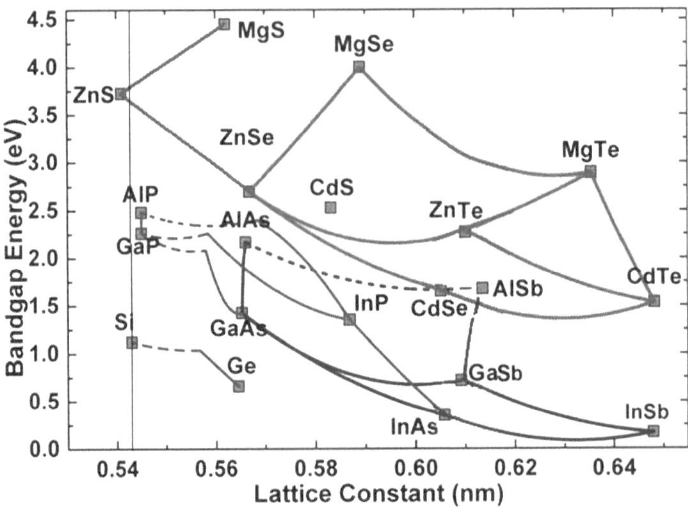

Fig. 2.9 Bandgap energy versus lattice constant [39]

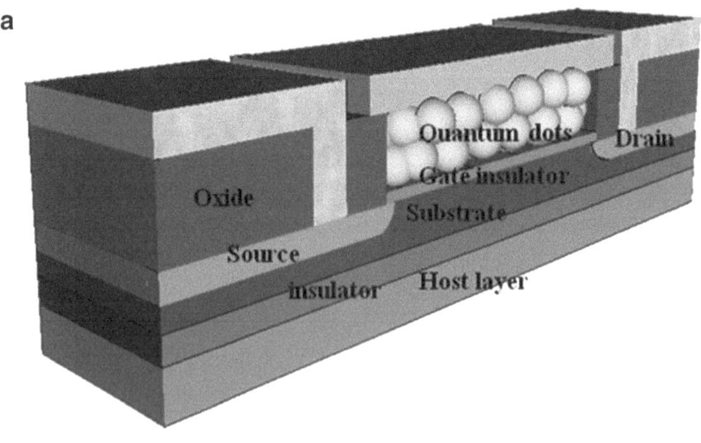

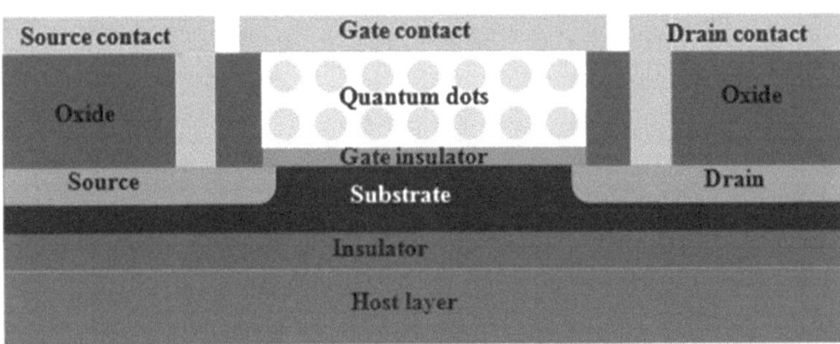

Fig. 2.10 (**a**) Device structure and (**b**) cross-sectional schematic of a QDGFET on silicon-on-insulator (SOI) substrate

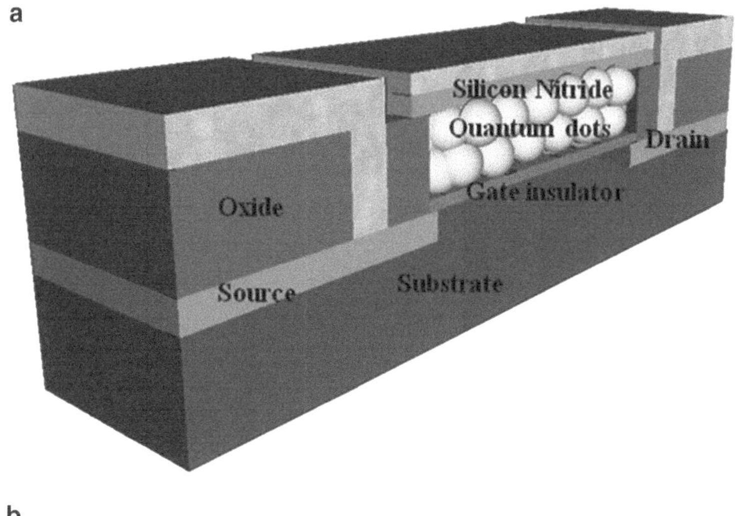

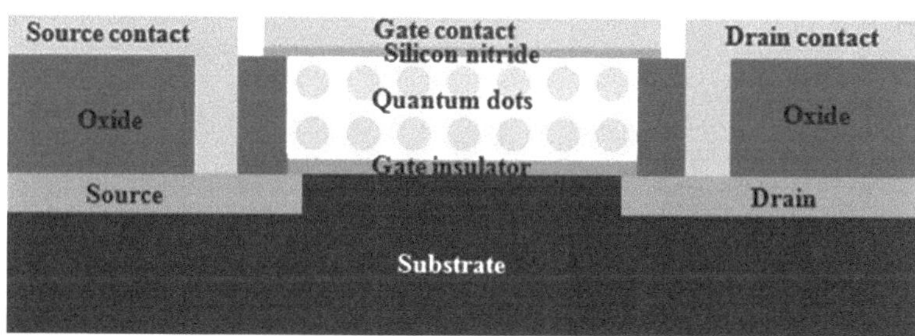

Fig. 2.11 (**a**) Device structure (**b**) cross-sectional schematic of QDGFET having thin film on top of quantum dots in the gate region

2.4.1.4 Thin Layer of SiN on Top of SiO_x-Cladded Si Dots on Top of SiO_2 Gate Insulator

The problem for ternary logic implementation is the noise margin which is affected by the stability of the intermediate state between on and off state of the device. In our previous work, we have shown three states of QDGFET with different types of quantum dots (QDs) on top of the gate insulator. One major problem of QDGFET is the stability of self-assembled QDs in the gate region which destroy the three-state behavior of the QDGFET with time. In this paper, we introduce an improved QDGFET structure which increases the stability of the QDs in the gate region as well as the QDGFET behavior. This improved structured QDGFET shows distinct three-state behavior even after 8–10 months after fabrication [40].

Device Structure: Device structure of a conventional QDGFET is shown in Fig. 2.11a where two layers of QDs are deposited on top of the gate insulator.

The improved device structure is shown in Fig. 2.11b where a thin layer of silicon nitride (SiN) layer having thickness 20 Å is deposited on top of the QDs.

Because of no control gate insulator in the conventional QDGFET, the quantum dots in the gate region are directly attached with the gate contact metal layer and are affected at the time of annealing during the fabrication process. In this improved structure, a 20-Å SiN layer is deposited on top of the QDs in the gate region (Fig. 1.1b). The thin insulator layer makes a shield for metal diffusion within the dots in the gate region. Beside this, the nitride film also prevents the reduction of the cladding layers of QDs and makes them stable.

References

1. Sah, Chih-Tang: Evolution of the MOS transistor – from conception to VLSI. Proc. IEEE **76**(10), 1280–1326 (1988)
2. Ihantola, H.K.J., Moll, J.L.: Design theory of a surface field-effect transistor. Solid-State Electron. **7**(4), 423–430 (1964)
3. Barron, M.B.: Low level currents in insulated gate field effect transistors. Solid-State Electron. **15**(3), 293–302 (1972)
4. Stuart, R.A., Eccleston, W.: Leakage currents of MOS devices under surface-depletion conditions. Electron. Lett. **8**(9), 225–227 (1972)
5. Swanson, R.M., Meindl, J.D.: Ion-implanted complementary MOS transistors in low-voltage circuits. IEEE J. Solid-State Circuits **7**(2), 146–153 (1972)
6. Van Overstraeten, R.J., Declerck, G., Broux, G.L.: Inadequacy of the classical theory of the MOS transistor operating in weak inversion. IEEE Trans. Electron Devices **20**(12), 1150–1153 (1973)
7. Natarajan, S., et al.: A 32nm logic technology featuring 2nd-generation high-k + metal-gate transistors, enhanced channel strain and 0.171μm2 SRAM cell size in a 291Mb array. In: IEDM Technical Digest, pp. 941–943, Dec 2008
8. Thompson, S., et al.: A 90 nm logic technology featuring 50 nm strained silicon channel transistors, 7 layers of Cu interconnects, low k ILD, and 1 μm2 SRAM cell. In: IEDM Technical Digest, pp. 61–64, Dec 2002
9. Bai, P., et al.: A 65nm logic technology featuring 35nm gate lengths, enhanced channel strain, 8 Cu interconnect layers, low-k ILD and 0.57 μm2 SRAM cell. In: IEDM Technical Digest, pp. 657–660, Dec 2004
10. Auth, C., et al.: 45nm high-k + metal gate strain-enhanced transistors. In: 2008 Symposium on VLSI Technology, pp. 128–129, June 2008
11. Balestra, F., et al.: Double-gate silicon-on-insulator transistor with volume inversion: a new device with greatly enhanced performance. IEEE Electron Device Lett. **8**(9), 410–412 (1987)
12. Hisamoto, D., et al.: Impact of the vertical SOI 'DELTA' structure on planar device technology. IEEE Trans. Electron Devices **38**(6), 1419–1424 (1991)
13. Lim, H.K., Fossum, J.G.: Threshold voltage of thin-film Silicon-on-insulator (SOI) MOSFET's. IEEE Trans. Electron Devices **30**(10), 1244–1251 (1983)
14. Yeo, Y.-C., Ranade, P., Lu, Q., Lin, R., King, T.-J., Hu, C.: Effects of high-_dielectrics on the workfunctions of metal and silicon gates. In: VLSI Technology Digest, pp. 49–50, June 2001
15. Yeo, Y.-C., Lu, Q., Ranade, P., Takeuchi, H., Yang, K.J., Polishchuk, I., King, T.-J., Hu, C., Song, S.C., Luan, H.F., Kwong, D.-L.: Dual-metal gate CMOS technology with ultra-thin silicon nitride gate dielectric. IEEE Electron Device Lett. **22**, 227–229 (2001)

16. Lee, S.J., Luan, H.F., Bai, W.P., Lee, C.H., Jeon, T.S., Senzaki, Y., Roberts, D., Kwong, D.-L.: High quality ultra thin CVD HfO gate stack with poly-Si gate electrode. In: IEDM Technical Digest, pp. 31–34, Dec 2000
17. Lu, Q., Lin, R., Ranade, P., Yeo, Y.C., Meng, X., Takeuchi, H., King, T.-J., Hu, C., Luan, H., Lee, S., Bai, W., Lee, C.-H., Kwong, D.-L., Guo, X., Wang, X., Ma, T.-P.: Molybdenum metal gate MOS technology for post-SiO gate dielectrics. In: IEDM Technical Digest, pp. 641–644, Dec 2000
18. Mönch, W.: Electronic properties of ideal and interface-modified metal–semiconductor interfaces. J. Vac. Sci. Technol. B **14**, 2985–2993 (1996)
19. Kavalieros, J., et al.: Tri-gate transistor architecture with high-k gate dielectrics, metal gates and strain engineering. In: 2006 Symposium on VLSI Technology, pp. 50–51, June 2006
20. Velliantis, G., et al.: Gatestacks for scalable high-performance FinFETs. In: 2007 Symposium on VLSI Technology, pp. 681–684, June 2007
21. Kang, Y.C., et al.: Effects of film stress modulation using TiN metal gate on stress engineering and its impact on device characteristics in metal gate/high-dielectric SOI FinFETs. IEEE Electron Device Lett. **29**(5), 487–490 (2008)
22. Colinge, J.P., et al.: Silicon-on-insulator 'gate-all-around device'. In: IEDM Technical Digest, pp. 595–598, Dec 1990
23. Monfray, S., et al.: 50 nm-gate all around (GAA)-silicon on nothing (SON)-devices: a simple way to co-integration of GAA transistors within bulk MOSFET process. In: 2002 Symposium on VLSI Technology, pp. 108–109, June 2002
24. Larrieu, G. et al.: Low temperature implementation of dopant-segregated band edge metallic S/D junctions in thin-body SOI p-MOSFETs. In: IEDM Technical Digest, pp. 147–150, Dec 2007
25. Hisamoto, D., et al.: A fully depleted lean-channel transistor (DELTA)-a novel vertical ultrathin SOI MOSFET. IEEE Electron Device Lett. **11**(1), 36–38 (1990)
26. Martini, I., Kamp, M., Fischer, F., Worschech, L., Koeth, J., Forchel, A.: Fabrication of quantum point contacts and quantum dots by imprint lithography. Microelectron. Eng. **57–58**, 397–403 (2001)
27. Verma, V.B., Reddy, U., Dias, N.L., Bassett, K.P., Li, X., Coleman, J.J.: Patterned quantum dot molecule laser fabricated by electron beam lithography and wet chemical etching. IEEE J. Quantum Electron. **46**(12), 1827–1833 (2010)
28. Jung, S.I., Yun, I., Han, I.K., Cho, S.M., Lee, J.I.: Fabrication and optical properties of CdSe/ZnS core/shell quantum-dot multilayer film and hybrid organic/inorganic light-emitting diodes fabricated by using layer-by-layer assembly. J. Korean Phys. Soc. **52**(6), 1891–1894 (2008)
29. Tan, Z., Xu, J., hang, C., Zhu, T., Zhang, F., Hedrick, B., Pickering, S., Wu, J., Su, H., Gao, S., Wang, A.Y., Kimball, B., Ruzyllo, J., Dellas, N.S., Mohney, S.E.: Colloidal nanocrystal-based light-emitting diodes fabricated on plastic toward flexible quantum dot optoelectronics. J. Appl. Phys. **105**(3), 034312-1-5 (2009)
30. Fukui, T., Saito, H., Kasu, M., Ando, S.: MOCVD methods for fabricating GaAs quantum wires and quantum dots. J. Cryst. Growth **124**(1–4), 493–496 (1992)
31. Elarde, V.C., Yeoh, T.S., Rangarajan, R., Coleman, J.J.: Controlled fabrication of InGaAs quantum dots by selective area epitaxy MOCVD growth. J. Cryst. Growth **272**(1–4), 148–153 (2004)
32. Matsumura, N., Tai, E., Kimura, Y., Saito, T., Ohira, M., Saraie, J.: Self-assembling CdSe, ZnCdSe and CdTe quantum dots on ZnSe(100) epilayers. Jpn. J. Appl. Phys. **39**, 1104–1105 (2000)
33. Bimberg, D., Grundmann, M., Ledentsov, N.N., Ruvimov, S.S., Werner, P., Richter, U., Heydenreich, J., Ustinov, V.M., Kop'ev, P.S., Alferov, Z.H.I.: Self-organization processes in MBE-grown quantum dot structures. Thin Solid Films **267**(1–2), 32–36 (1995)
34. Tiwari, S., Rana, F., Chan, K., Hanafi, H., Chan, W., Buchanan, D.: Volatile and non-volatile memories in silicon with nano-crystal storage. In: IEDM, pp. 521–525, Dec 1995
35. Jain, F.C., Heller, E., Karmakar, S., Chandy, J.: Device and circuit modeling using novel 3-state quantum dot gate FETs. In: International Semiconductor Device Research Symposium, 12–15 Dec 2007, College Park

36. Gogna, M., Karmakar, S., Al-Amoody, F., Papadimitrakopoulous, F., Jain, F.: Self Assembled Germanium Oxide cladded Germanium quantum dot gate nonvolatile memory. In: Nanoelectronic Devices for Defense and Security, 28 Sep–02 Oct 2009, Fort Lauderdale
37. Phely-Bobin, T., Chattopadhyay, D., Papadimitrakopoulos, F.: Characterization of mechanically attrited Si/SiOx nanoparticles and their self-assembled composite films. Chem. Mater. **14**, 1030–1036 (2002)
38. Jain, F., Papadimitrakopoulos, F.: Site-specific nanoparticle self-assembly. US Patent 7,368,370, 2008
39. Karmakar, S., Suarez, E., Jain, F.: Quantum dot gate three state FETs using ZnS – ZnMgS lattice-matched gate insulator on silicon. J. Electron. Mater. **40**(8), 1749–1756 (2011)
40. Karmakar, S., Gogna, M., Jain, F.C.: Improved device structure of quantum dot gate FET to get more stable intermediate state. Electron. Lett. **48**(24), 1556–1557 (2012)

Chapter 3
Quantum Dot Gate Field-Effect Transistors: Fabrication and Characterization

This chapter discusses different fabrication steps and characterization results of QDGFETs. Characterization of different quantum dots as well as gate insulator layer is also presented in this chapter. Transfer characteristics and output characteristics of QDGFETs are also presented in this chapter. The improvement of subthreshold swing in SOI (silicon-on-insulator) is also demonstrated in this chapter.

3.1 Fabrication Methodology

3.1.1 GeO_x-Cladded Ge Quantum Dots on Top of High-κ Gate Dielectric

The quantum dot gate FET (QDGFET) is fabricated using four masks. Mask 1 is used to open the source and thc drain for n-type doping. Mask 2 is used to open the gate window between the source and the drain. Mask 3 is used to open the source and the drain contact window, and mask 4 is used to define different contact pads.

3.1.1.1 Cleaning of Silicon Wafer

1. A piece of p-type (100) silicon wafer (resistivity 10 Ω-cm) was first cleaned using boiling trichloroethylene (TCE) for 5 min to remove different types of contamination such as epoxy or silicon resins [1].
2. The wafer was then cleaned using boiling acetone for 5 min to remove TCE.
3. This process was followed by placing the sample in boiling methanol for 5 min to remove acetone.
4. The wafer then rinsed in deionized (DI) water thoroughly to remove any trace of methanol and blow-dry with N_2 jet.
5. Then the sample was dipped in 5 % hydrofluoric acid (HF) solution for 5 min to remove any native oxide.

S. Karmakar, *Novel Three-state Quantum Dot Gate Field Effect Transistor: Fabrication, Modeling and Applications*, DOI 10.1007/978-81-322-1635-3_3,

6. The sample was then rinsed with DI water thoroughly.
7. Then the sample was dipped in piranha solution (1:1 sulfuric acid to hydrogen peroxide) for 5 min. Silicon dioxide grew on the sample in this process [2].
8. The sample was then rinsed thoroughly in DI water.
9. Dipped in 5 % hydrofluoric acid (HF) solution for 5 min to remove oxide.
10. The sample was then dipped in DI water to clean thoroughly.
11. Soaked the sample in freshly made RCA solution (6 H_2O:1 H_2O_2:1HCl) at 70 °C for 5 min.
12. Rinsed in deionized (DI) water.
13. Soak in HF for 5 min.
14. The sample was then transferred to boiling propanol for transferring to wet oxidation furnace.

3.1.1.2 Source and Drain Processing

1. Wet oxide was grown at 1,000 °C for 15 min which grew 1,700 Å oxide on top of the sample (Fig. 3.1B). This wet oxide acts as a mask for phosphorous diffusion in the source-drain region and also provides good visibility of different patterns for the following photolithography process.
2. A 2-in. mounting wafer was placed on a spinner and spun with S1813 positive photoresist at 1,000 rpm for 10 s.
3. The sample was then placed in the center of the mounting wafer and placed on hot plate at 115 °C for 5–6 min until the sample adhered to the wafer.
4. The mounting wafer was then placed on the spinner and spin-coated the sample with S1813 positive photoresist at 5,000 rpm for 30 s (Fig. 3.1C).
5. The mounting wafer with the sample was then baked (preexpose bake) at 115 °C for 2 min.
6. Exposed the sample in UV for 30 s with mask 1 (Fig. 3.1D).
7. Pre-develop baked at 115 °C for 1 min.
8. Developed the sample in 3.5:1 (water to 351 Developer) developer solution for 10–15 s (Fig. 3.1E).
9. Rinsed thoroughly in DI water to clean developer solution and post-baked the sample for 10 min at 115 °C.
10. Etched field oxide grown by wet oxidation process from source and drain region using buffered oxide etch for few minutes and checked whether the oxide had been removed (Fig. 3.1F).
11. Removed sample from mounting wafer using acetone.
12. Cleaned the photoresist using acetone. Then cleaned with methanol and DI water.
13. Placed the sample in boiling propanol for transferring it to phosphorous diffusion furnace [3, 4].
14. The sample was dried off and placed on a boat with phosphorous source. The sample was then loaded in phosphorous diffusion furnace. Phosphorous diffusion [5, 6] was done at 1,000 °C for 5 min in a nitrogen (N_2) environment (Fig. 3.1G).

3.1.1.3 Gate Processing

The source-drain diffusion is followed by gate processing:

1. Steps 2–12 of the source-drain processing were performed using mask 2 (gate opening mask) (Fig. 3.1H–J).
2. Sample was stored in boiling isopropyl alcohol and transferred to dry oxidation furnace for sacrificial gate oxide growth [7, 8].
3. The sample was then placed into an oxidation furnace in an oxygen environment at 900 °C for 20 min and 1,050 °C for 10 min. This oxidation process grows approximately 400-Å gate oxide (Fig. 3.1K).

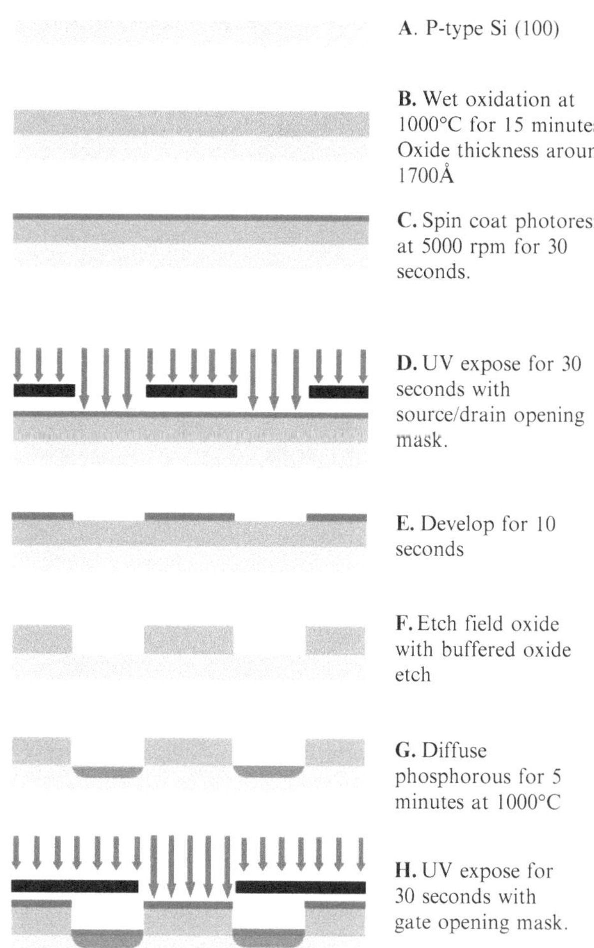

Fig. 3.1 Process flow of QDGFET fabrication

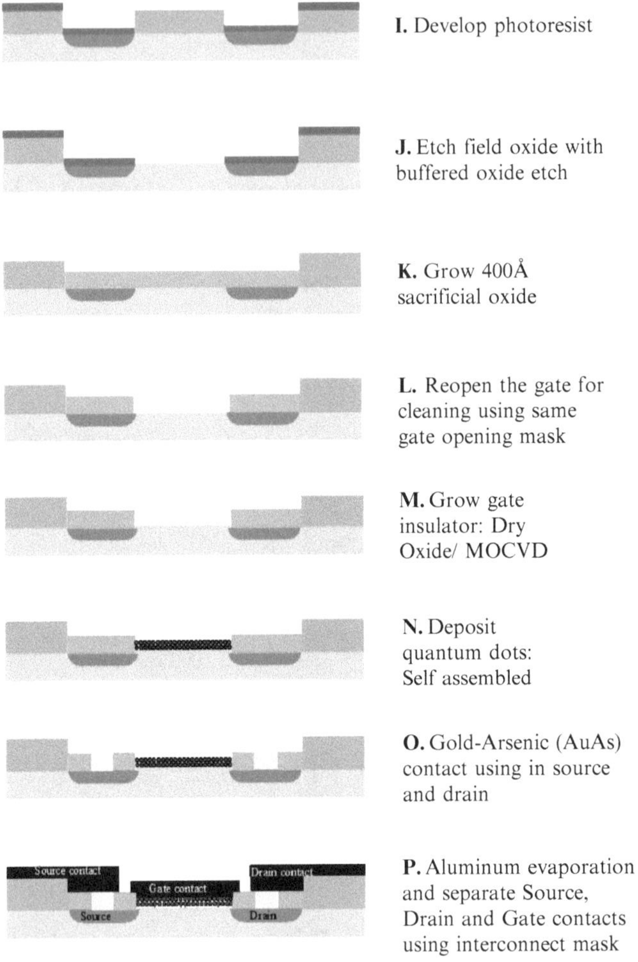

Fig. 3.1 (continued)

Reopening Gate Region

The sacrificial oxide grown in the previous step was to clean the gate region. The growth of sacrificial oxide was followed by reopening the gate region to grow tunnel insulator (Fig. 3.1L):

1. Steps 2–12 of the source-drain processing were performed.
2. Sample was stored in boiling isopropyl alcohol and transferred to metal-organic chemical vapor deposition reactor for II–VI insulator deposition.

Deposition of II–VI Insulator

The reopening of gate region is followed by the deposition of II–VI gate insulator layers. This is accomplished by MOCVD [9].

1. After reopening the gate region, the sample was transferred to MOCVD reactor.
2. The silicon under the gate region was cleaned by the hydrogen sulfide (H_2S) at 800 °C for 30 min before the actual growth.
3. A ZnS buffer layer was grown at 477 °C for 1 min using the flow of metal-organic source dimethylzinc (DMZn) at 20 sccm and diethyl sulfide (DES) at 50 sccm.
4. The growth of ZnS buffer layer was followed by the growth of high bandgap ZnMgS layer. This layer was grown at 333 °C with 35 mW/cm^2 UV radiation for 2 min. The different metal-organic sources were dimethylzinc (DMZn) at 20 sccm, diethyl sulfide (DES) at 50 sccm, and bismethylcyclopentadienyl magnesium [$(MeCp)_2$ Mg].
5. The high bandgap ZnMgS layer was finally capped with a ZnS layer grown at 333 °C for 1 min using the same metal-organic source flow rate as in step 3 (Fig. 3.1M).

Germanium Dot Self-Assembly

The growth of lattice-matched insulator layer was followed by the site-specific self-assembly of GeO_x-Ge quantum dots [10].

1. Germanium nanoparticles were mixed with an oxidative agent and sonicated for 2 days in a sealed flask. The pH of the solution was maintained at 5.5.
2. The solution was then centrifuged and etched for three-dimensional consignments. Self-assembly was then conducted for 3 min. Because of the pH of the dot solution, the dots are positively charged because of ionization [11, 12] and deposited only on the *p*-doped channel between the *n*-doped source and drain regions (Fig. 3.1N).

3.1.1.4 Source-Drain Contact Processing

After gate processing, the next step was to make source-drain contact.

1. The sample was cleaned using acetone and gently blow-dry.
2. Steps 2–9 of the source-drain processing were performed using mask 3 (source-drain contact hole opening mask).
3. After post bake, the source-drain contact is etched using buffered oxide etch to remove GeO_x-cladded Ge dots from the source and drain.
4. Ammonium hydroxide-to-hydrogen peroxide ($NH_4OH{:}H_2O_2 = 1{:}4$) solution was used to etch the II–VI layer underneath the quantum dot layer on top of the source-drain region. The etch rate was around 30 Å/s. The grown II–VI layer thickness was about 75–80 Å. Etching was done for around 20–30 s until the silicon underneath was exposed.
5. The sample was cleaned in DI water thoroughly.
6. AuAs was evaporated on top of the sample to form source-drain contact. Before loading the sample inside the evaporator, the sample was dipped into the buffered oxide etch solution for 5 s to remove any native oxide grown during the transferring time to evaporator.

7. Photoresist was cleaned using acetone and forms source-drain contact by liftoff method (Fig. 3.1O).
8. The sample was then annealed at 300 °C in a nitrogen environment. In case of germanium dots in the gate region, forming gas was not used because germanium dots are unstable and their GeO_x cladding decomposes due to presence of hydrogen in the forming gas. Although the eutectic temperature of Au is 375 °C, germanium dots decompose at a temperature above 350 °C.
9. After annealing, gate contact was formed by evaporating 1,000-Å aluminum all over the sample.

3.1.1.5 Interconnect Formation

This is the last processing step. In this process, the source-drain-gate contact pads were patterned.

1. After aluminum evaporation, the sample was gently blow-dry to remove any dust on it.
2. Steps 2–9 of the source and drain processing were performed.
3. After post bake, the aluminum was etched using the aluminum etchant.
4. The photoresist was cleaned using acetone, and the sample was ready for testing (Fig. 3.1P).

3.1.2 *QDGFET: SiO_x-Cladded Si Quantum Dots on Silicon-on-Insulator Substrate*

For silicon-on-insulator (SOI) substrate, the cleaning method and phosphorous diffusion method were same as the silicon substrate. The source and drain diffusion was followed by the gate insulator deposition process. In this process, the gate was opened using gate opening mask, and a sacrificial oxide of thickness 400 Å was grown on top of gate region using dry oxidation method at 900 °C for 20 min and 1,050 °C for 10 min. This sacrificial oxide was used for cleaning the gate region. The sacrificial oxide in the gate region was cleaned by reopening the gate region using buffered oxide etch solution. The tunneling silicon dioxide insulator growth was followed by the reopening of the gate region. Dry oxidation at 800 °C for 5 min grew around 20-Å gate oxide for the QDGFET.

3.1.2.1 SiO_x-Cladded Si Dot Self-Assembly

The SiO_X-cladded Si quantum dots were site-specifically self-assembled on top of the tunnel oxide in the gate region. Silicon nanoparticles were ball-milled first for 12 h to get around 40–50-nm nanoparticle size. Then they were sonicated for 2 days in a sealed flask with an oxidative agent. The solution was then centrifuged and etched for three-dimensional consignments. In this process, the pH of the solution was maintained within 4.5 and 5.5. The dots are positively charged due to ionization of the acidic environment of the solution. The sample was then dipped in the dot solution for 3 min. Because of the presence of n-type interface charge in the p-type silicon substrates, these positively charged dots assembled on top of the tunneling oxide in the gate region. Self-assembly occurs because the dots cannot deposit on top of the n-type source and drain regions and only deposit on top of the p-type channel in between the n-type source and drain regions.

Source and drain contacts were formed by opening the source and drain contact holes and etching the deposited silicon dioxide and quantum dot layers on top of the source and drain regions. Gold arsenic was thermally evaporated on top of the sample, and source-drain contact deposition was done by the liftoff method. The final ohmic contact was formed by annealing the contacts at 375 °C for 1 min in a nitrogen-hydrogen (N_2:H_2 = 9:1) environment. Since silicon dots are stable up to 750 °C, source-drain contact can be annealed at eutectic temperature of gold. Finally, aluminum evaporation was done to form the gate contact and source-drain contact pads.

3.2 Quantum Dot Gate Characterization

This section describes the atomic force microscopy (AFM) and transmission electron microscopy (TEM) techniques that have been used to characterize the fabricated FETs and circuits.

3.2.1 *Atomic Force Microscopy (AFM)*

AFM [13, 14] provides a 3D profile of the surface on a nanoscale, by measuring forces between a sharp probe (<10 nm) and surface at a very short distance (0.2–10-nm probe-sample separation).

The probe is supported on a flexible cantilever. The AFM tip "gently" touches the surface and records the small force between the probe and the surface [15, 16].

The probe is placed on the end of a cantilever (which one can think of as a spring). The amount of force between the probe and sample depends on the spring constant of the cantilever and the distance between the probe and the sample surface.

The motion of the probe across the surface is controlled similarly to the STM using feedback loop and piezoelectric scanners. The primary difference in instrumentation design is how the forces between the probe and sample surface are monitored. The deflection of the probe is typically measured by a "beam bounce" method. A semiconductor diode laser is bounced off the back of the cantilever onto a position-sensitive photodiode detector. This detector measures the bending of cantilever, while the tip is scanned over the sample. The measured cantilever deflections are used to generate a map of the surface topography. Figure 3.2 shows the experimental setup for an AFM system. Figures 3.3 and 3.4 show the

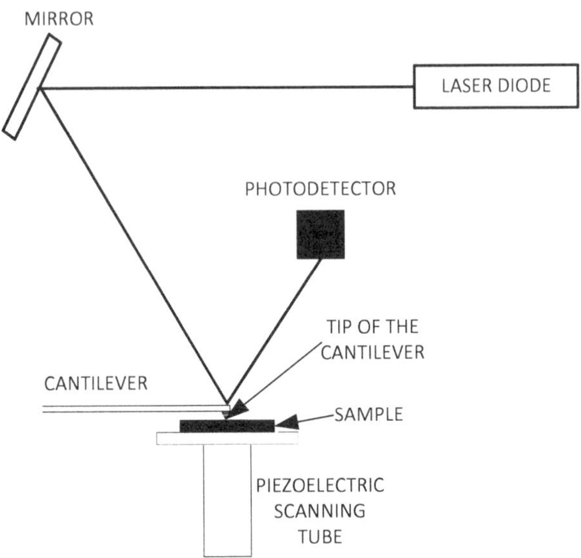

Fig. 3.2 Experimental setup for atomic force microscopy (AFM)

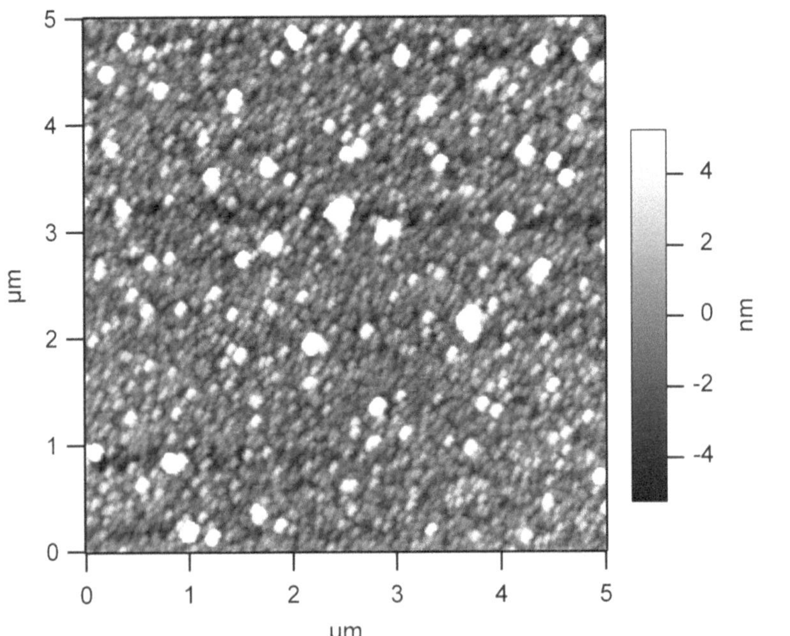

Fig. 3.3 The atomic force microscopy (AFM) image indicates the deposition of silicon quantum dots having 4-nm size

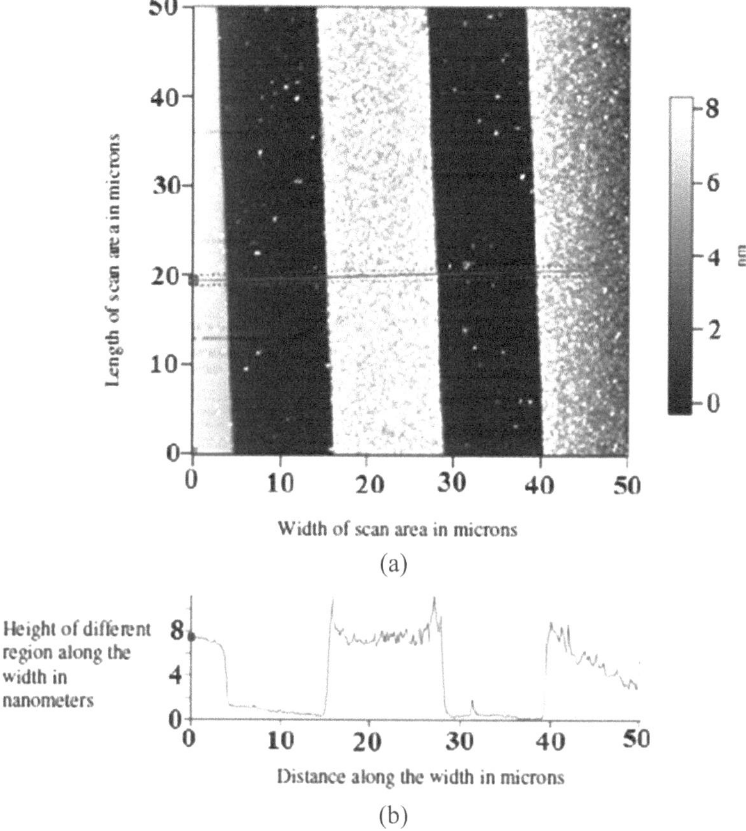

Fig. 3.4 Atomic force microscopy (AFM) image shows the deposition of GeOx-cladded Ge dots on patterned silicon wafer [10]

atomic force microscopy (AFM) image of deposited SiO_x-cladded Si dots and GeO_x-cladded Ge quantum dots on top of silicon substrate.

3.2.2 Transmission Electron Microscopy (TEM)

In TEM, an electron gun shines a beam of monochromatic electrons. This stream is focused to a small, thin, coherent beam by the use of two condenser lenses [17]. The first lens (usually controlled by the "spot size knob") largely determines the "spot size," the general size range of the final spot that strikes the sample. The second lens (usually controlled by the "intensity or brightness knob") actually changes the size of the spot on the sample, changing it from a wide dispersed spot to a pinpoint beam.

The beam is restricted by the condenser aperture (usually user selectable), knocking out high-angle electrons. The beam strikes the specimen, and parts of it are transmitted. This transmitted portion is focused by the objective lens into an image. Optional objective and selected area metal apertures can restrict the beam; the objective aperture enhances contrast by blocking out high-angle diffracted electrons, and the selected area aperture enables the user to examine the periodic diffraction of electrons by ordered arrangements of atoms in the sample.

The image is passed down the column through the intermediate and projector lenses, being enlarged all the way to the screen.

The image strikes the phosphor image screen, and light is generated, allowing the user to see the image. The darker areas of the image represent those areas of the sample that fewer electrons were transmitted through (they are thicker or denser). The lighter areas of the image represent those areas of the sample that more electrons were transmitted through (they are thinner or less dense).

The schematic diagram of a TEM system is shown in Fig. 3.5. Figures 3.6 and 3.7 show the TEM image of isolated silicon dots and isolated germanium dots, respectively.

3.3 ZnS Layer Characterization

This section describes the characterization of ZnS layer grown on silicon substrate by metal-organic chemical vapor deposition (MOCVD) technique.

3.3.1 X-Ray Diffraction

Diffraction occurs as waves interact with a regular structure having repeat distance about the same as the wavelength. Light can be diffracted by a grating having scribed lines spaced on the order of a few thousand angstroms, about the wavelength of light.

It happens that X-rays [18] have wavelengths on the order of a few angstroms, the same as typical interatomic distances in crystalline solids. Crystals are ordered, three-dimensional arrangements of atoms with characteristic periodicities. As the spacing between atoms is on the same order as X-ray [19, 20] wavelengths (1–3 Å), crystals can diffract the radiation when the diffracted beams are in phase.

In 1912, W. L. Bragg recognized a predictable relationship among several factors. Constructive interference only occurs for certain angle (θ's) correlating to a (*hkl*) plane. Specifically when the path difference is equal to n wavelengths where n is an integer, the Bragg equation is given as $n\lambda = 2d\sin\theta$ where d is the spacing between two crystal planes. For a given wavelength (λ), diffraction [21] can only occur at a certain angle (θ) for a given d-spacing. The wavelength of the X-ray (λ) in this case is 1.54 Å.

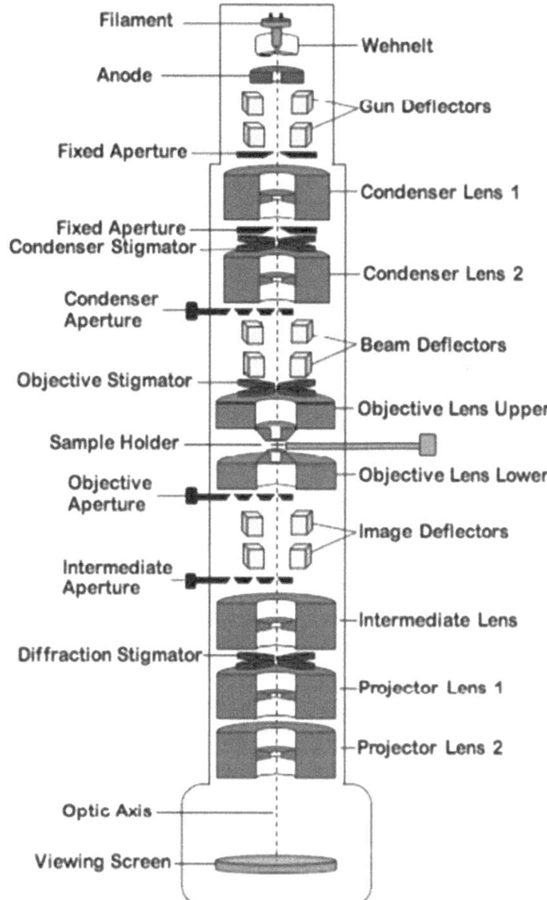

Fig. 3.5 Schematic diagram of a TEM lens system

Figure 3.8 shows the X-ray diffraction from crystal lattices. Figure 3.9 [22] shows the X-ray diffraction data from a ZnS thick film grown on silicon by MOCVD method. This data shows the growth of single-crystalline ZnS on silicon which is nearly lattice matched.

3.4 Cross-Sectional High-Resolution Transmission Electron Micrograph (HRTEM)

The sample for cross-sectional TEM was prepared by a FEI Strata 400S Dual-Beam FIB instrument. A 1-μm thick layer of platinum (Pt) was deposited onto the surface of the FET to minimize the ion beam damage to the device during FIB machining. Then a 1-μm thick membrane having a dimension of 10 μm across and 5 μm in height was

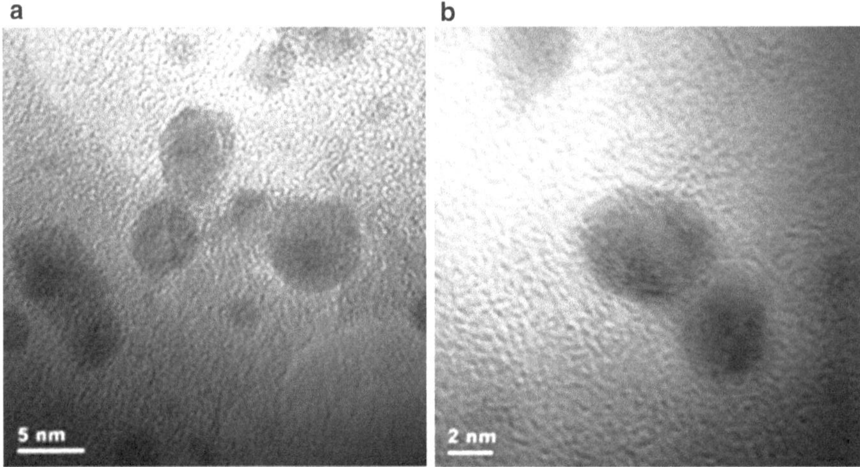

Fig. 3.6 Transmission electron microscopy image of Si nanoparticles after etching and oxidizing with an average particle size (APS) of 6 nm [23]

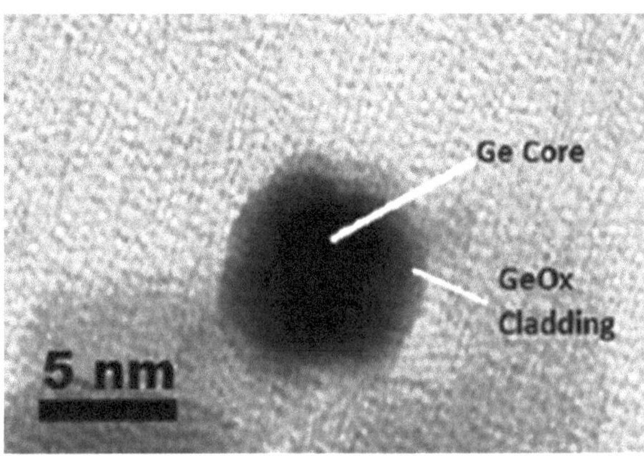

Fig. 3.7 Transmission electron microscopy image of germanium nanoparticles after etching and oxidizing with an average particle size (APS) of 4 nm [10]

cut through the Pt layer using Ga + ion beam. An omniprobe micromanipulator was then used to relocate the membrane to the flip stage. Finally, the membrane was welded to a FIB TEM grid on the stage using Pt. Further Ga + ion milling was then performed until the membrane was thin enough for TEM observation.

A JEOL 2010 FasTEM, operating at 200 kV, was used to examine the TEM specimens. This instrument is equipped with a high-resolution objective lens pole piece (spherical aberration coefficient Cs = 0.5 mm), giving a point-to-point resolution of <0.19 nm in phase-contrast images. Chemical microanalysis was performed in situ using an EDAX Phoenix atmospheric thin window EDXS.

TEM image (Fig. 3.10) obtained from the experiment shows the deposition quantum dots in different QDGFET structures.

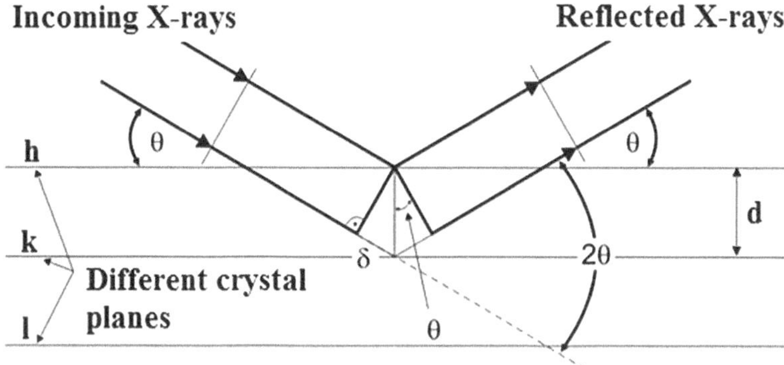

Fig. 3.8 X-ray diffraction from crystal planes

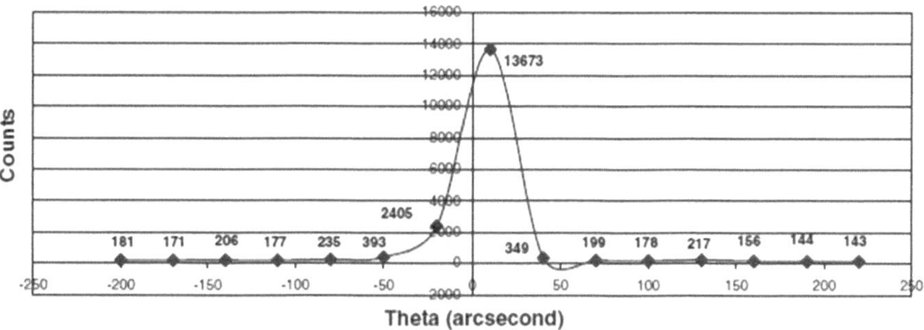

Fig. 3.9 X-ray diffraction data from ZnS thin film [22]

3.5 Electrical Characteristics

This section presents the electrical characteristics (transfer characteristic and output characteristic) of different fabricated QDGFETs. The transfer characteristic shows the variation of drain current with respect to gate voltage where drain-to-source voltage is used as a parameter. In the output characteristic, the variation of drain current with the variation of drain-to-source voltage is examined where gate-to-source voltage is used as a parameter.

3.5.1 GeOx-Cladded Ge Dots on Top of ZnS-ZnMgS Gate Insulator

Figure 3.11 shows the measured transfer characteristics ($I_D - V_{GS}$) of a fabricated QDGFET having a 15-μm channel width and a 5-μm channel length. Note the observation of an intermediate state "i" between the on

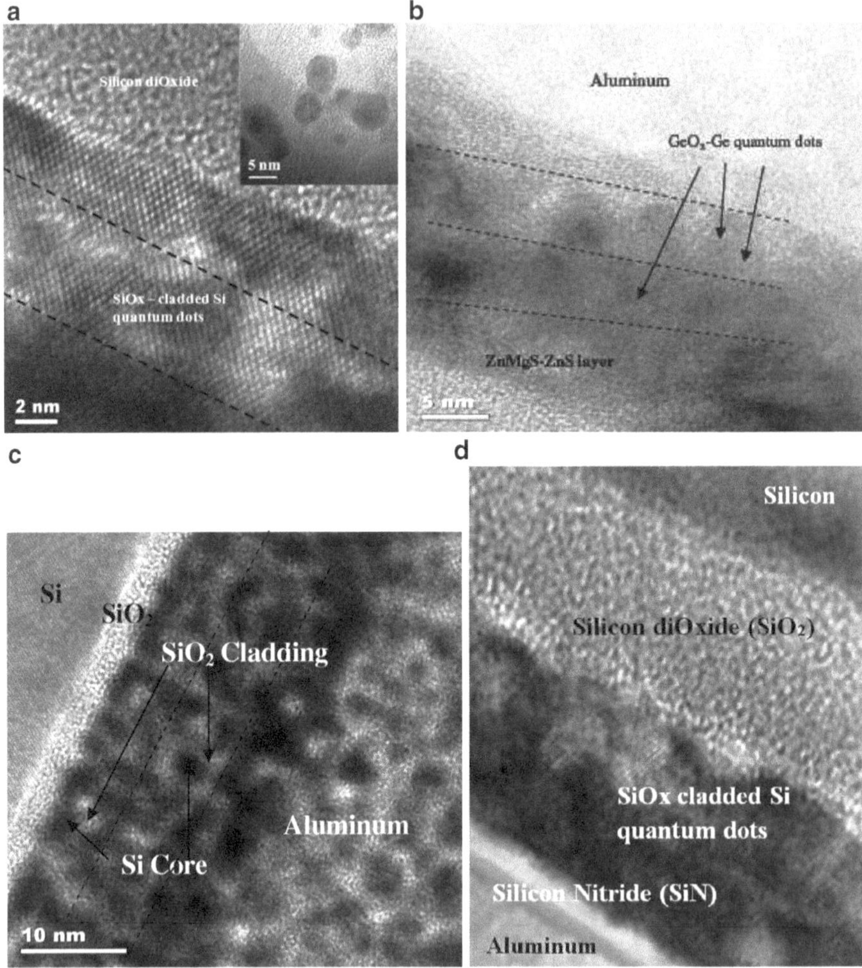

Fig. 3.10 High-resolution transmission electron micrograph (HETEM) image of QDGFET: (**a**) self-assembled SiO_x-cladded Si dots deposition on top of SiO_2 tunnel gate insulator, (inset) TEM image of self-assembled SiO_x-cladded Si dots, (**b**) self-assembled GeO_x-cladded Ge dots on top of lattice-matched ZnS-ZnMgS gate insulator on the p-Si substrate [10], (**c**) self-assembled SiO_x-cladded Si dots deposition on top of SiO_2 tunnel gate insulator of a QDGFET on silicon-on-insulator (SOI) substrate, and (**d**) silicon nitride layer on top of SiO_x-cladded Si dots in improved QDGFET structure

(>2.5 V) and the off (<0.2 V) states of the transistor due to the charge transfer from the inversion channel to the Ge quantum dots (which are ~4 nm in diameter having 1-nm GeO_x cladding). The intermediate state is a low-current saturation state at a drain current of 300 nA for gate voltages (V_{GS}) between 0.5 and 1.0 V. Figure 3.12 shows the output characteristics

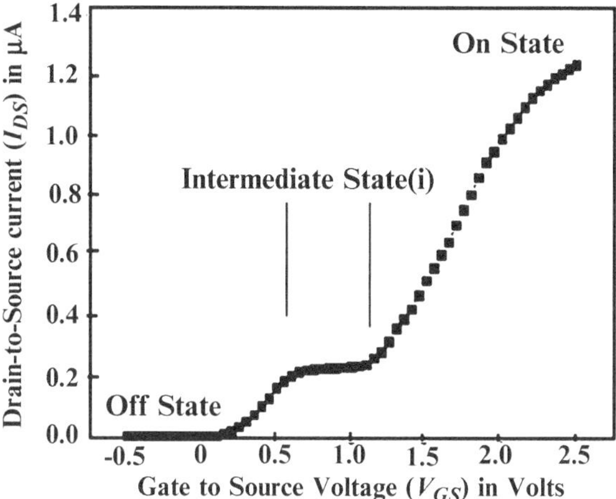

Fig. 3.11 Transfer characteristics (I_D–V_{GS}) of a fabricated QDGFET having 15-μm channel width and 5-μm channel length

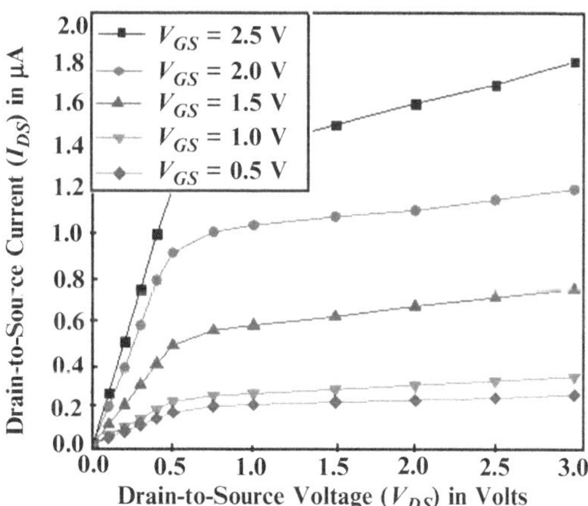

Fig. 3.12 Output characteristics (I_D–V_{DS}) of the QDGFET. The I_D changes very little with V_{DS} when V_G is between 0.5 and 1.0 V

($I_D - V_{DS}$) of the quantum dot gate FET. The I_D changes very little with V_{DS} when V_{GS} is between 0.5 and 1.0 V.

3.5.2 SiO_x-Cladded Si Dots on Top of SiO_2 Tunnel Insulator in SOI Substrate

The transfer characteristic of the fabricated QDGFET on SOI substrate having 15-μm channel width (W) and 5-μm channel length (L) is shown in Fig. 3.13. In

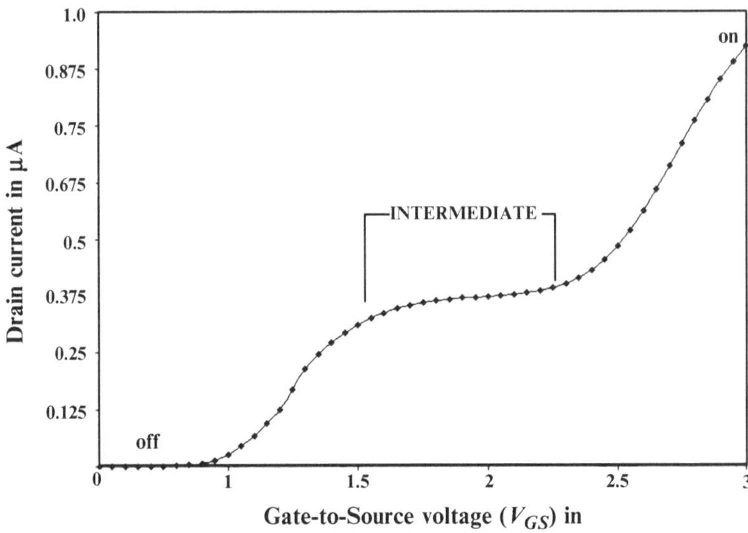

Fig. 3.13 Transfer characteristic of the fabricated QDGFET on SOI substrate

this experiment, a constant drain-to-source voltage of 0.5 V was applied across the QDGFET, and the gate voltage varied from 0 to 3 V. When the gate voltage varies, the underneath substrate of the gate region changes gradually from accumulation to depletion to inversion channel region. The application of constant drain-to-source voltage drags electrons from source-to-drain region through the inversion channel underneath the gate. The applied gate-to-source voltage corresponding to the formation of inversion channel is known as threshold voltage which is around 1.1 V for this device. When the applied gate voltage is more than the threshold voltage, electrons tunnel from the inversion channel to different quantum dot layers on top of the gate region. The stored electrons in the quantum dot layers in the gate region oppose the applied gate voltage, and effective gate voltage on the inversion channel remains almost constant. This effect produces the intermediate state in the transfer characteristics of the QDGFET. This intermediate state is a low-current saturation state at a drain current of 375 μA for gate voltage (V_{GS}) between 1.5 and 2.4 V. As the gate voltage increases (beyond 2.4 V), two quantum dot layers on top of the gate region become saturated with the electrons tunneling from the inversion channel, and they cannot hold more electrons. So the effective applied gate voltage increases again, and the drain-to-source current also increases. Figure 3.14 shows the logarithmic plot of the transfer characteristic in the subthreshold regime. The improvement in subthreshold slope in the subthreshold regime is because of the SOI substrate. Figure 3.15 shows the output characteristics (I_D vs. V_{DS}) of the fabricated QDGFET. The I_D changes very little with V_{DS} when V_{GS} is in the intermediate range.

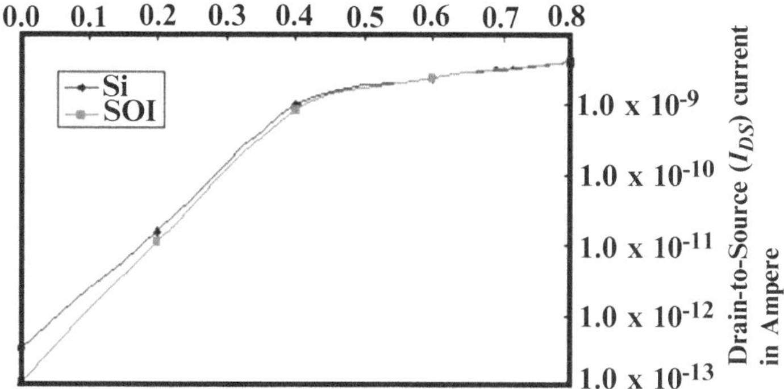

Fig. 3.14 Logarithmic plot of the transfer characteristic in subthreshold regime

Fig. 3.15 Output characteristics (I_D vs. V_{DS}) of the fabricated QDGFET

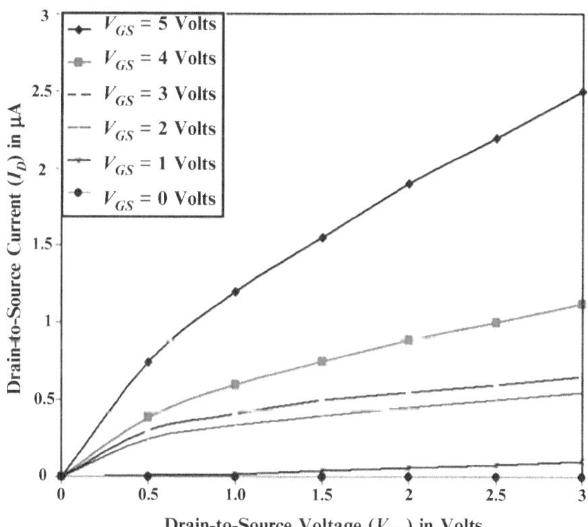

3.5.3 *Thin Layer of Silicon Nitride on Top of SiO_x-Cladded Si Quantum Dots in the Gate Region of FET*

The experimental transfer characteristics (drain current vs. gate voltage) of a conventional QDGFET and the improved QDGFET are shown in Fig. 3.16a, b, respectively. The experimental data shows that the charge storage capability in the conventional QDGFET structure decreases with time and the intermediate state gradually disappears with time. But the intermediate state remains almost same in the new improved QDGFET structure even after 10 months.

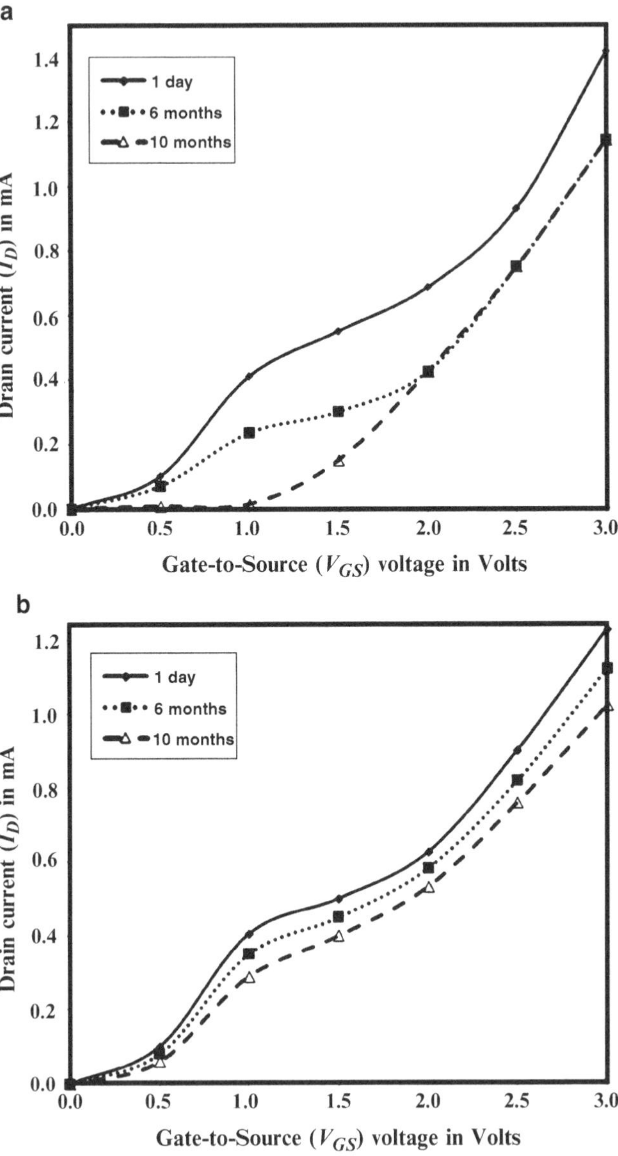

Fig. 3.16 Experimental transfer (I_D vs. V_{GS}) characteristics of (**a**) conventional QDGFET and (**b**) new improved QDGFET

The preliminary data shows the stable intermediate state in the new improved version of QDGFET for a longer time than the conventional QDGFET. The three-state nature of conventional QDGFET is less stable because of breaking of cladding layer of the deposited QDs in the gate region. The breaking of cladding layer increases the charge leakage between different QDs in the gate region and destroys the three-state behavior of the QDGFET. The presence of SiN layer in the improved device structure protects the cladding layer of the QDs in the gate region, and three-state behavior of the device remains for a longer period. The new improved design of QDGFET will help to develop better ternary logic circuit in future [23].

References

1. Kern, W.A., Poutinen, D.A.: The measurement of effective complex refractive indices for selected metal silicides. RCA Rev. **31**, 187 (1970)
2. dos Santos Filho, S.G., Hasenack, C.M., Salay, L.C., Mertens, P.: A less critical cleaning procedure for silicon wafer using diluted HF dip and boiling in isopropyl alcohol as final steps. J. Electrochem. Soc. **142**(3), 902–907 (1995)
3. Gandhi, S.K.: The Theory and Practice of Microelectronics. Wiley, New York (1968)
4. Crank, J.: The Mathematics of Diffusion. Oxford University Press, Walton Street, Oxford (1956)
5. Grove, A.S.: Physics and Technology of Semiconductor Devices. Wiley, New York (1967)
6. Fair, R.B.: On the role of self-interstitials in impurity diffusion in silicon. J. Appl. Phys. **51**, 5828 (1980)
7. Doremus, R.H.: Oxidation of silicon by water and oxygen and diffusion in fused silica. J. Phys. Chem. **80**(16), 1773–1775 (1976)
8. Sze, S.M.: Physics of Semiconductor Devices, 2nd edn. Wiley, New York (1981)
9. Jain, F.C., Suarez, E., Gogna, M., AlAmoody, F., Butkiewicus, D., Hohner, R., Liaskas, T., Karmakar, S., Chan, P.Y., Miller, B., Chandy, J., Heller, E.: Novel quantum dot gate FETs and nonvolatile memories using lattice-matched II-VI gate insulators. J. Electron. Mater. **38**(8), 1574–1578 (2009)
10. Karmakar, S., Suarez, E., Jain, F.: Quantum dot gate three state FETs using ZnS – ZnMgS lattice-matched gate insulator on silicon. J. Electron. Mater. **40**(8), 1749–1756 (2011)
11. Phely-Bobin, T., Chattopadhyay, D., Papadimitrakopoulos, F.: Characterization of mechanically attrited Si/SiOx nanoparticles and their self-assembled composite films. Chem. Mater. **14**, 1030–1036 (2002)
12. Jain, F., Papadimitrakopoulos, F.: Site-specific nanoparticle self-assembly. US Patent 7,368,370, 2008
13. Occelli, M.L., Gould, S.A.C.: The use of atomic force microscopy (AFM) to study the surface topography of commercial fluid cracking catalysts (FCCs) and pillared interlayered clay (PILC) catalysts. In: Studies in Surface Science and Catalysis. Proceedings of the American Chemical Society Petroleum Division Conference: Fluid Catalytic Cracking VI, Philadelphia, Pennsylvania, USA, **149**, 71–104 (2004)
14. Schiraldi, D.A., Occelli, M.L., Gould, S.A.C.: Applications of atomic force microscopy to current problems in industrial polyester chemistry. Polym. News **27**(6), 195–200 (2002)
15. Schiraldi, D.A., Occelli, M.L., Gould, S.A.C.: Atomic force microscopy (AFM) study of poly (ethylene terephthalate-co-4, 4′-bibenzoate): a polymer of intermediate structure. J. Appl. Polym. Sci. **82**(11), 2616–2623 (2001)
16. Occelli, M.L., Gould, S.A.C.: Examination of coked surfaces of pillared rectorite catalysts with the atomic force microscope. J. Catal. **198**(1), 41–46 (2001)
17. Fultz, B., Howe, J.M.: Transmission Electron Microscopy and Diffractometry of Materials, 3rd ed., Springer, Berlin Heidelberg New York (2008). Corr. 2nd printing, 2008
18. Warren, B.E.: X-ray Diffraction. General, Dover Publications Inc., New York (1969/1990)

19. Cullity, B.D.: Elements of X-ray Diffraction, 2nd edn. Addison-Wesley, Reading (1978)
20. Als-Nielsen, J., McMorrow, D.: Elements of Modern X-ray Physics. Wiley, New York (2001)
21. Bowen, D.K., Tanner, B.K.: High Resolution X-ray Diffractometry and Topography. Taylor & Francis, London/Bristol (1998)
22. Suarez, E., Gogna, M., Al-Amoody, F., Karmakar, S., Ayers, J., Heller, E., Jain, F.: Nonvolatile memories using quantum dot (QD) floating gate assembled on II–VI tunnel insulator. J. Electron. Mater. **39**(7), 903–907 (2010)
23. Karmakar, S., Gogna, M., Jain, F.C.: Improved device structure of quantum dot gate FET to get more stable intermediate state. Electron. Lett. **48**(24), 1556–1557 (2012)

Chapter 4
Quantum Dot Gate Field-Effect Transistors: Theory and Device Modeling

The device model of the QDGFET and its theory are discussed in this chapter. The modification of band diagram of QDGFET due to the presence of quantum dots in the gate region is shown in this chapter. The theory of operation based on self-consistent solution of Schrödinger and Poisson equations is also presented in this chapter.

4.1 Band Diagram of a MOSFET

The band diagram [1–5] of an ideal metal-oxide-semiconductor field-effect transistor is shown in Fig. 4.1. In an ideal MOS diode

1. The flat band voltage $= V_{FB} = \Phi_m - \Phi_s$ where Φ_m is the work function of metal and Φs is the work function of the semiconductor.
2. $V_{FB} = \Phi_m - \Phi_s = \Phi_m - [\chi + E_g/(2q) + \Psi_B] = 0$ for p-type substrate where χ = electron affinity in semiconductor, E_g = energy bandgap of silicon, and Ψ_B = potential difference between the Fermi level (E_f) and intrinsic level (E_i) in bulk Si.
3. The charges are at the gate and at the channel. These are equal and opposite. There is no charge in the oxide or at the interface of SiO_2/Si.
4. No carrier transport through the oxide layer.

In a real device, because of the presence of different surface-trapped charges, the band diagram deviates from this ideal nature. In a real MOS diode, flat band voltage is not zero. There is always a difference between the metal work function and the semiconductor work function. In equilibrium condition, the Fermi level of gate metal and the semiconductor channel Fermi level should be aligned. Different parameters such as the metal work function, electron affinity, and Fermi levels are shown in the abovementioned figure.

S. Karmakar, *Novel Three-state Quantum Dot Gate Field Effect Transistor: Fabrication, Modeling and Applications*, DOI 10.1007/978-81-322-1635-3_4,

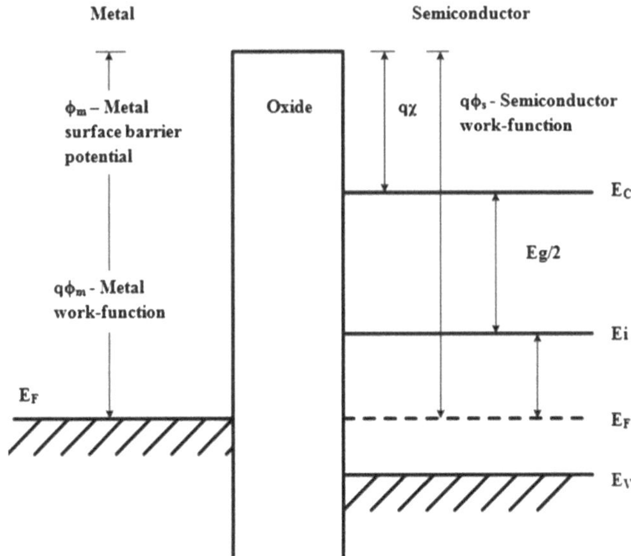

Fig. 4.1 Band diagram of an ideal MOS structure

4.2 Theory of Operations of a MOSFET

When we apply an external voltage to the metal gate, based on the polarity and magnitude of the gate voltage, the following mode of operation happens within the MOSFET.

4.2.1 Accumulation

A negative voltage at the gate metal will attract some free holes to the interface. An increased number of free holes are created at the semiconductor-insulator interface. Since there is an accumulation of majority carriers in the interface, this mode is known as accumulation. Figure 4.2 shows the accumulation of majority carriers (p type) in the interface.

4.2.2 Strong Accumulation

For stronger negative voltages at the gate, the Fermi level is forced below the valence band at the interface. A channel with a high density of free holes is created. This is called strong accumulation.

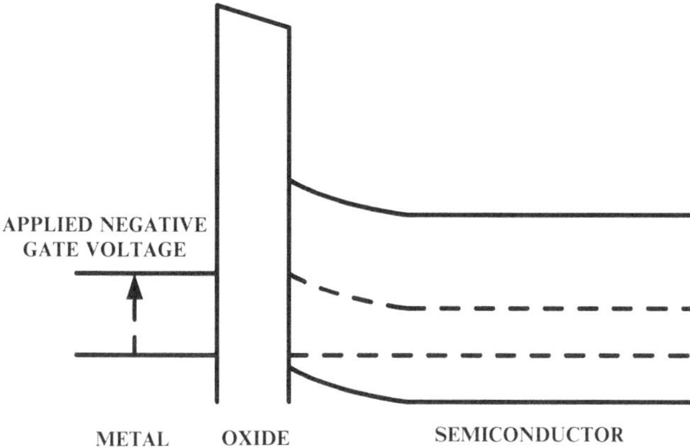

Fig. 4.2 Accumulation of majority carriers (p type) for negative gate voltage

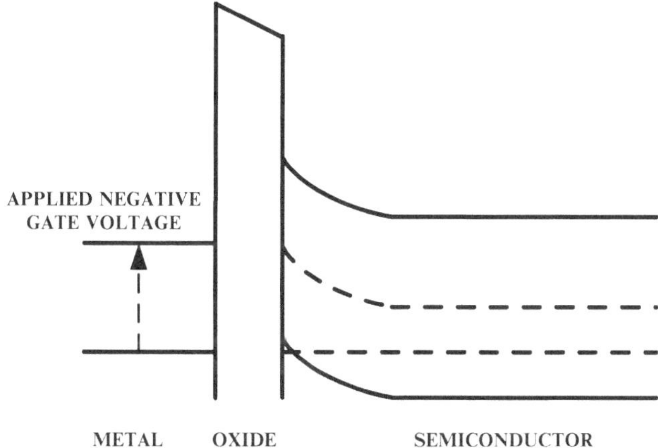

Fig. 4.3 Accumulation of majority carriers for high negative gate voltage

Again, as for the strong inversion case, further increase in the voltage will result in an increase of the density of free holes rather than an increase of the channel in space. Figure 4.3 shows the accumulation of majority carriers for high negative gate voltage.

4.2.3 Depletion

On the other hand, when a small positive voltage is applied to the metal gate, free holes are pushed out of the interface region, and band bending results on the other side of the insulator. Free carriers (holes) flow out of the interface region, and a depletion region is formed. The uncompensated (negatively) ionized acceptors cause an electric field and a parabolic bending of the bands.

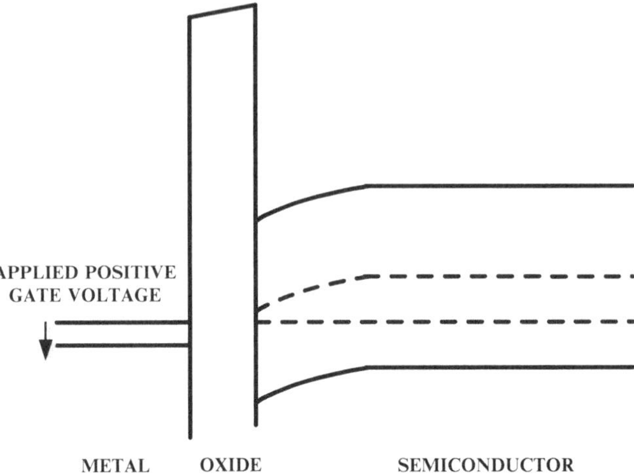

Fig. 4.4 Depletion region formation for small value of positive gate voltage

There is also a voltage drop in the oxide. In the oxide, no charge can reside, and hence the field is constant and the voltage drop is linear in space. The total voltage drop plus the band bending is equal to the external voltage. Figure 4.4 shows the depletion region formation for a small positive gate voltage.

4.2.4 *Weak Inversion*

When the gate voltage is further increased at the interface, an inversion region is created. The semiconductor surface near the oxide-semiconductor interface becomes n type, although not very conductive yet. Figure 4.5 shows the weak inversion condition for small positive gate voltage.

4.2.5 *Strong Inversion*

When the gate voltage becomes more and more negative, the Fermi level crosses the conduction band close to the interface. This is called strong inversion. Free electrons are in the so-called channel next to the oxide. This channel is therefore highly conductive.

The more availability of states in the conduction band means that further increase in the gate voltage will not extend this strong inversion region into space, but rather will increase the density of electrons in the channels. The channel is always infinitesimally thin. The huge amount of free carriers can easily cause a large voltage drop (band bending), and only a thin layer is needed to absorb the external voltage (Fig. 4.6).

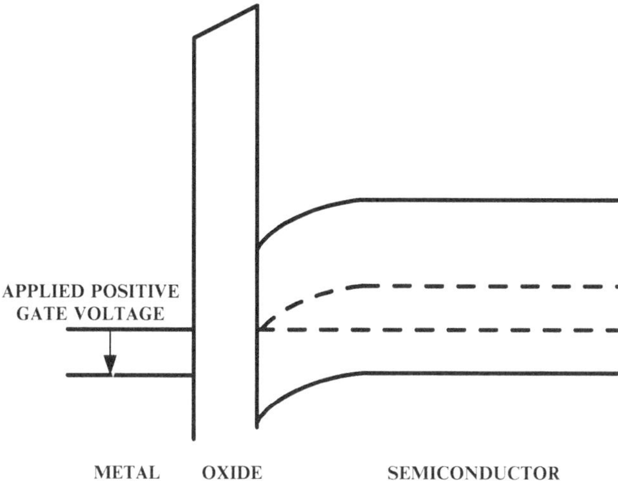

Fig. 4.5 Weak inversion condition for small positive voltage in gate

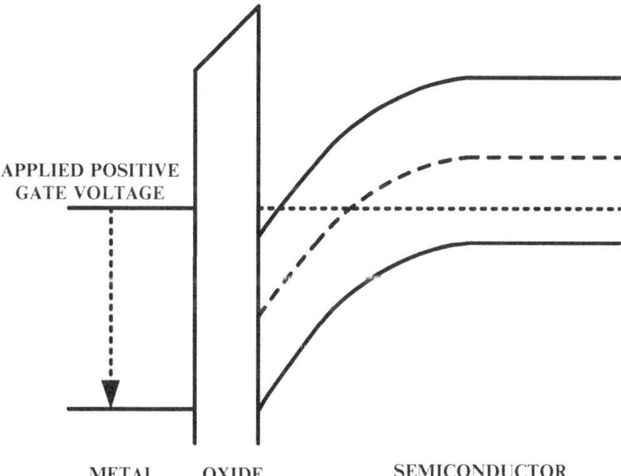

Fig. 4.6 Inversion channel formation

4.3 Band Diagram of a QDGFET

The energy band diagram of a QDGFET having SiO_x-cladded Si dots on top of SiO_2 gate insulator is shown in Fig. 4.7. The cladded quantum dot gate is formed by site-specific self-assembly (SSA) [6] of monodispersed cladded quantum dots, which preferentially deposit on the *p*-doped channel region.

Figure 4.7 shows the energy band diagram of a quantum dot gate field-effect transistor (QDGFET). The two quantum dot layers are shown with their cores

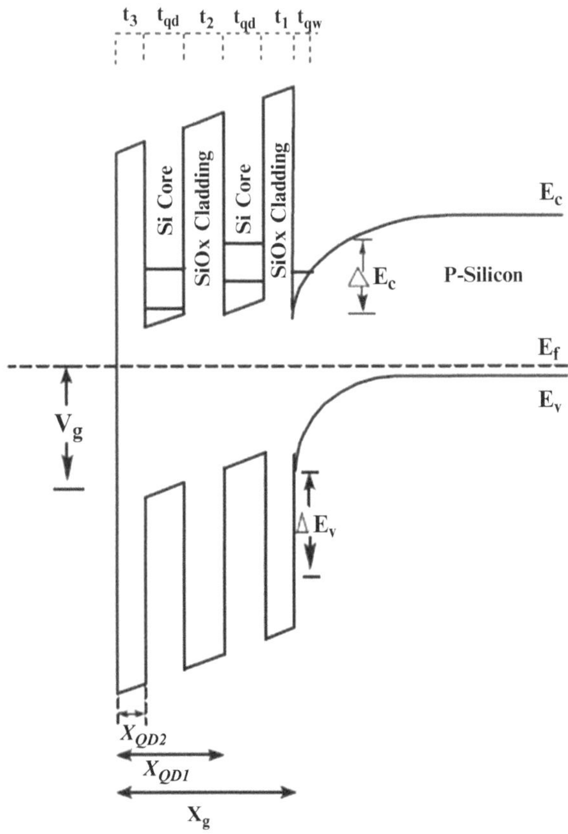

Fig. 4.7 Energy band diagram of a QDGFET having SiOx-cladded Si quantum dots on top of SiO_2 gate insulator

labeled as t_{qd}: the layer t_1 includes SiO_x cladding (1 nm) and any tunneling insulator (~2 nm), layer t_2 thickness is 2 nm (comprising of the two SiO_x claddings 1 nm each), and t_3 is the outer layer (~1 nm) which represents the cladding SiO_x thickness of outer layer of dots. In addition, the thickness of the quantum well inversion layer is represented by t_{qw}. Energy levels in the dots (whose location depends on dot parameters as well the electric field) are also shown.

4.4 Theory of Operations of QDGFET

4.4.1 SiO_x-Cladded Si Quantum Dots on SiO_2 Gate Insulator

Figure 4.8a–d shows the energy band diagrams when gate voltage is applied. The energy band diagram is calculated by solving Schrödinger and Poisson equations self-consistently following the procedure reported by our group [7], under the condition of inversion. Finally, the solution is determined for a range of gate voltages, which

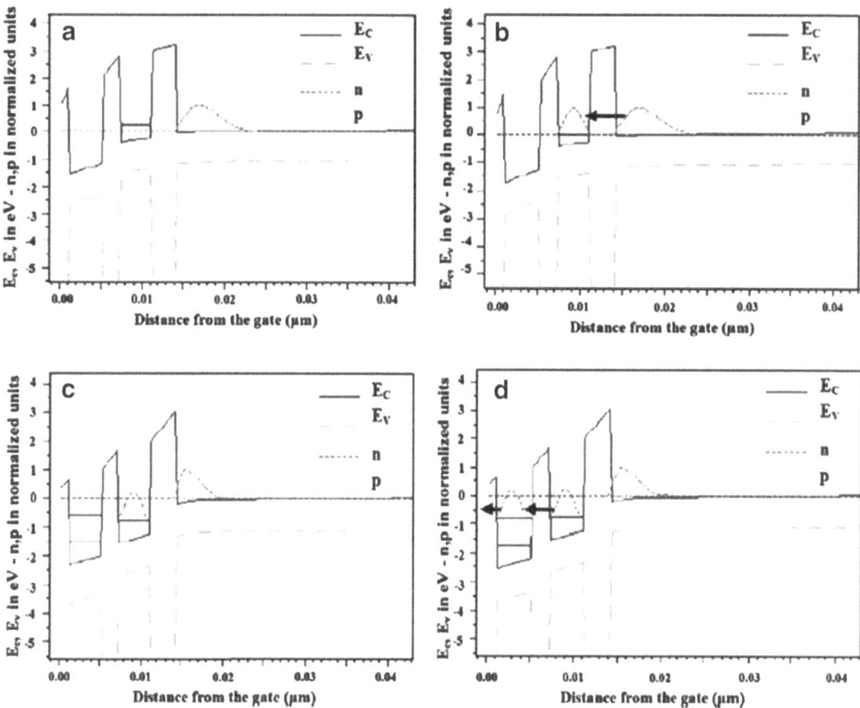

Fig. 4.8 (**a–d**) Energy band diagrams when gate voltage increases from (**a**) to (**d**) gradually

provides the charge in each quantum dot as a function of bias. These values are used to determine the shift in the threshold voltage (ΔV_{TH}) as the device is driven. The parameters used in the simulation are shown in Table 4.1. The wave function profile, measure of carrier concentration, is shown in Fig. 4.8. As the gate voltage is increased well above the threshold voltage, electrons are first transferred to the first quantum dot layer near the channel. This is due to the fact that the levels are lower in the inner QD layer. As the gate voltage is further increased, the charges are transferred from the first quantum dot layer to the second quantum dot layer near the gate electrode.

The observation of an intermediate state in the quantum dot gate FET can be explained by the electron tunneling from the inversion channel to either layer of quantum dots in the gate region. This charge sharing between channel and different quantum dot layers depends on the tunneling probability of the wave functions of the quantum well channel and quantum dots.

The charge on a dot in a layer is calculated by determining the tunneling rate of transition. The tunneling transition rate from the channel to the quantum dot layers, $P_{w \to d}$, is expressed by Hamiltonian in Eq. 4.1 following Chuang et al. [8–12].

$$P_{w \to d} = \frac{4\pi}{\hbar} \sum_{w,d} |\langle \psi_d | H_t | \psi_w \rangle|^2 (f_w - f_d) \delta(E_d - E_w) \quad (4.1)$$

Table 4.1 Device parameters for charge control simulation

Channel length (L)	5 μm
Channel width (W)	5 μm
QD gate capacitance (C_{OX})	4.4×10^{-7} F/cm^2
Drain-source voltage (V_{DS})	3 V
Mobility (μ_n)	600 cm^2/V.s
Si QD diameter	4 nm
SiOx barrier height	3.25 eV
Threshold voltage without QD gate charge (V_{THO})	0.6 V
Total gate layer	~14 nm
Effective dielectric constant	8
Center of QD1 from gate	~9 nm
Center of QD2 from gate	~3 nm

Here, ψ_d and ψ_w are the wave functions in the quantum dot and inversion layers, respectively, f_w and f_d are the Fermi distribution functions, H_t is the Hamiltonian, E_d and E_w are the energy levels in the inversion channel and in the quantum dots, and $\hbar$ is the reduced Planck's constant.

The tunneling [13–16] is responsible for the transfer of charge from the channel to the quantum dot layers. Note that in the absence of any charge transfer, there is some charge in the gate due to the presence of QD layers. This charge determines the threshold voltage. Using equations for conventional MOS device, the threshold voltage can be expressed as

$$\begin{aligned} V_{TH} &= V_{FB} + \frac{1}{C_{ox}} qN_A \sqrt{\frac{2\varepsilon_{sr}\varepsilon_o(2\psi_B + V_x)}{qN_A}} + 2\psi_B, \\ V_{FB} &= \varphi_{ms} - \frac{Q_{OX}}{C_{OX}}, \\ \psi_B &= \frac{kT}{q} \ln \frac{N_A}{n_i} \end{aligned} \tag{4.2}$$

The change in threshold voltage V_{TH} depends on surface potential ψ, metal and channel work function difference φ_{ms} and oxide charge Q_{ox}, and oxide capacitance C_{ox} for a given doping N_A in p-Si. The charge is distributed at the Si dot-SiO$_x$ interface near the channel.

The change in the flat band voltage which is also the threshold voltage shift due to electron tunneling can be expressed by Eq. 4.3:

$$\begin{aligned} \Delta V_{FB} &= \Delta V_{TH} \\ &= -\frac{q}{C_{ox}} \int_0^{x_g} \frac{x\rho(x)}{x_g} dx = -\frac{q}{C_{ox}} \left[\sum \frac{x_{QD1} n_1 N_{QD1}}{x_g} + \sum \frac{x_{QD2} n_2 N_{QD2}}{x_g} \right] \end{aligned} \tag{4.3}$$

Here, x_{QD1} and x_{QD2} are the distances of quantum dot cores from the gate contact; n_1 and n_2 are the number of dots in layer 1 and 2, respectively; N_{QD1} and

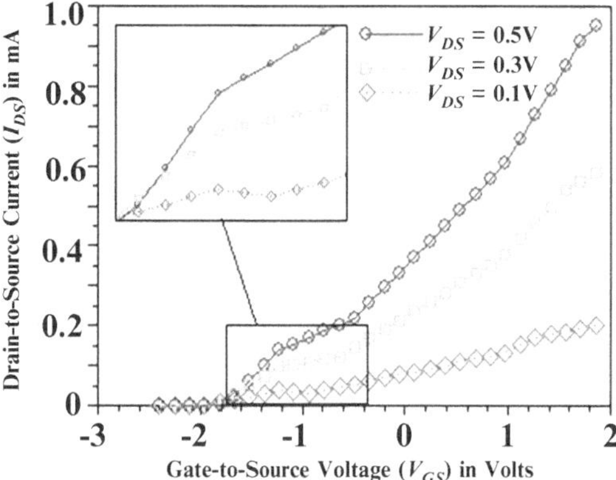

Fig. 4.9 Device simulation characteristics which show the generation of intermediate states between on and off states of a QDGFET [17]

N_{QD2} are the charges on each SiO_x-cladded Si quantum dots; and C_{ox} is the oxide capacitance. In addition, x_g is the distance of the Si-SiO_X interface from the gate, ρ is the charge density, and q is the electron charge. The effect of voltage shift is weighted according to the location of the charge, i.e., the closer to the oxide-semiconductor interface, the more shift it will cause.

The threshold voltage increases with the increase in gate voltage because of charge tunneling from the channel to quantum dots in the gate region. The drain current is expressed as Eq. 4.4:

$$I_D = \left(\frac{W}{L}\right) C_{ox}\mu_n \left[(V_{GS} - V_{TH})V_{DS} - \frac{V_{DS}^2}{2}\right] \quad (4.4)$$

The variation in the threshold voltage causes little increase in the drain current, because the ($V_{GS\_}V_{TH}$) term remains constant. The increase in gate voltage is compensated by the increase in the threshold voltage, and the drain current almost remains constant which produce the intermediate state of the device. The range of gate voltages in which this phenomenon will arise depends on the relative location of the quantum dot energy levels with respect to the energy levels of the quantum well inversion channel. Figure 4.9 shows the device simulation characteristics which shows the generation of intermediate state between on and off states of a QDGFET.

A similar solution can be obtained for SOI substrate having SiO_x-cladded Si quantum dots on top of SiO_2 gate insulator.

4.4.2 GeO_x-Cladded Ge Quantum Dots on ZnS-ZnMgS Gate Insulator

Figure 4.10 shows the energy band diagram of a quantum dot gate field-effect transistor (QDGFET). The two quantum dot layers are shown with their core labeled as t_{qd}, the layer t_1 includes GeO_x cladding (1 nm) and any tunneling insulator (~2 nm), layer t_2 thickness is 2 nm (comprising of the two GeO_x claddings

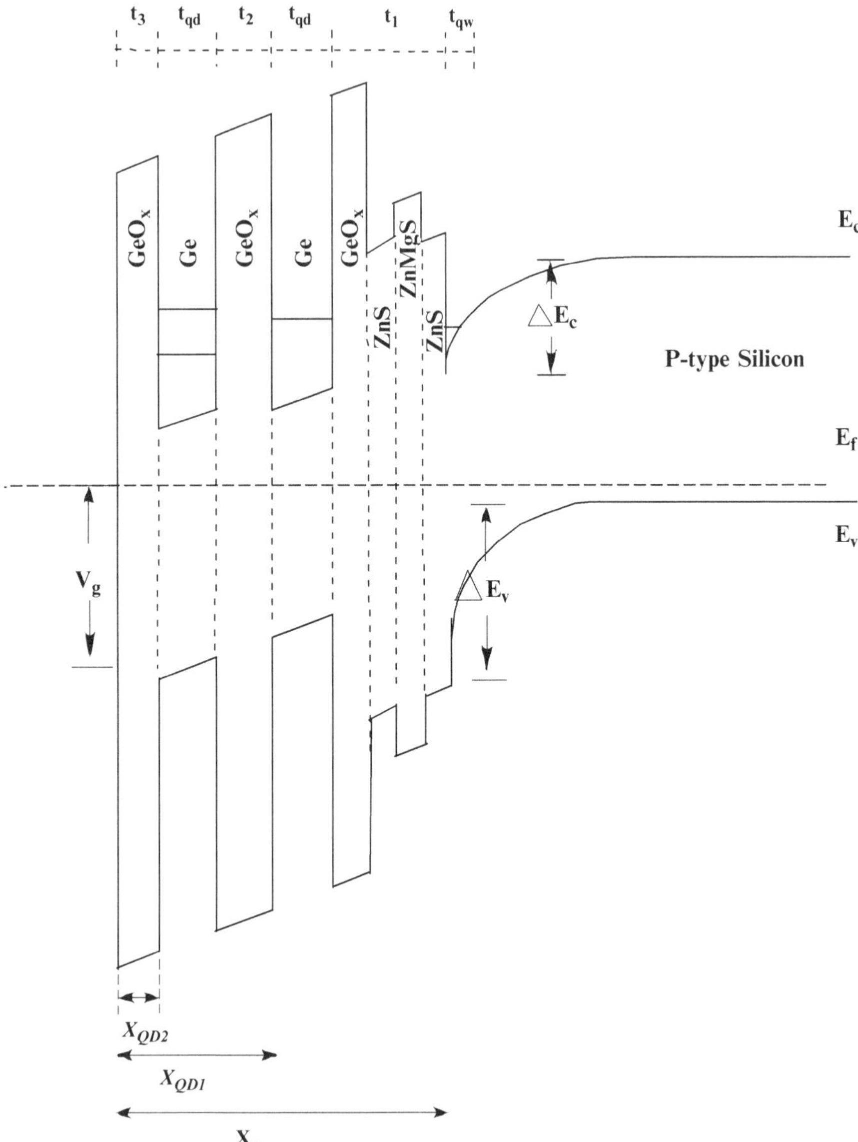

Fig. 4.10 Energy band diagram of a QDGFET having GeOx-cladded Ge quantum dots on top of ZnS-ZnMgS gate insulator

1 nm each), and t_3 is the outer layer (~1 nm) which represents the cladding GeO_x thickness of outer layer of dots. In addition, the thickness of the quantum well inversion layer is represented by t_{qw}. Energy levels in the dots (whose location depends on dot parameters as well the electric field) are also shown.

Figure 4.11a–d shows energy band diagrams and carrier concentration plots in the quantum well channel (~14 nm from gate contact) and two quantum dot gate

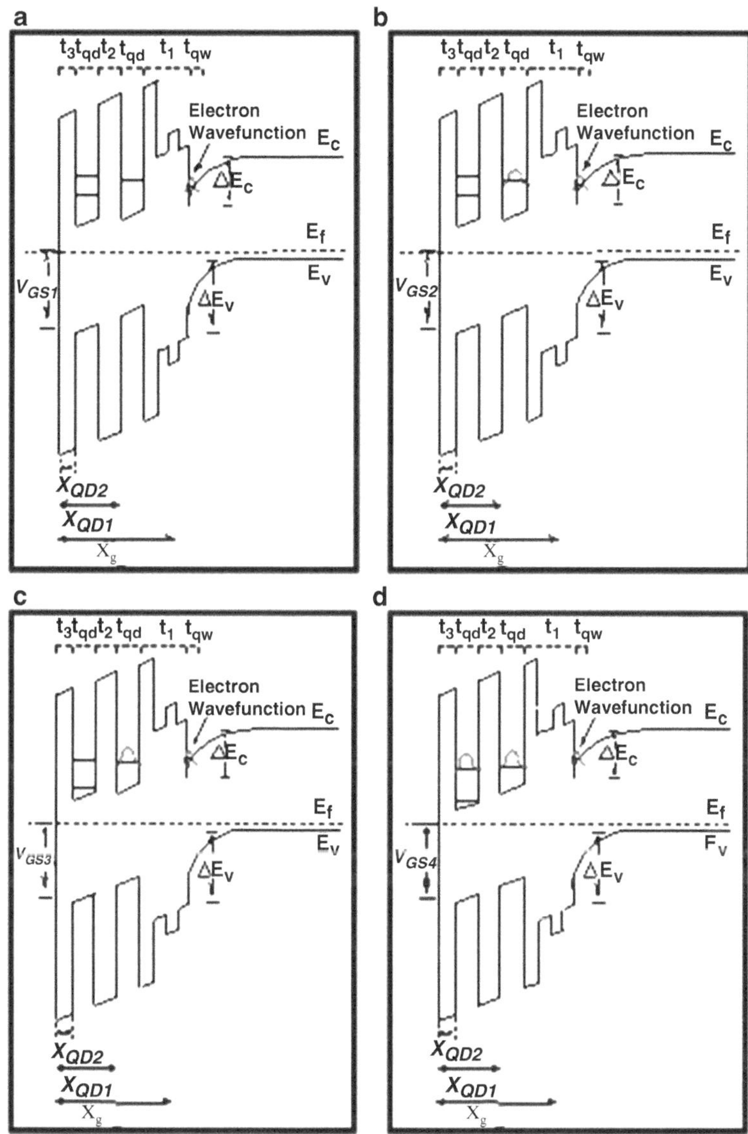

Fig. 4.11 (**a–d**) Energy band diagrams when gate voltage increases from (**a**) to (**d**) gradually

layers (~3 nm and ~9 nm from gate) at various gate voltages. Figure 4.11a shows the state of the device at a gate voltage before the dots are charged, Fig. 4.11b the charging of the first layer of dots closest to the channel via tunneling, Fig. 4.11c the buildup of the charge in QD layer 1 as the energy levels in the second QD layer are not lined up, and Fig. 4.11d the resonant tunneling to QD layer 2 (due to alignment of the bound states in QD1 and QD2) and discharging to the gate.

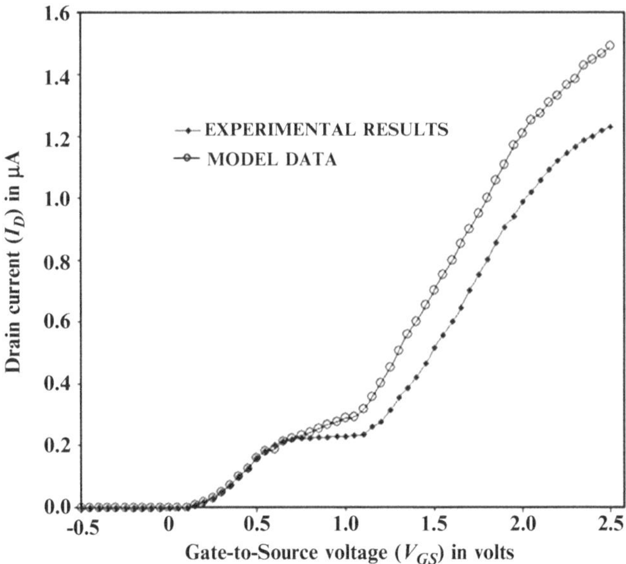

Fig. 4.12 Comparison of simulation results with experimental data for germanium dot QDGFET

Same self-consistent model that solves Schrödinger and Poisson equations with built-in transfer of carriers from the inversion channel to two layers of cladded GeOx-Ge quantum dots is used to explain the "i" state. Figure 4.12 shows simulated $I_D - V_{GS}$ curves for 0.5 V V_{DS}.

The simulation (Fig. 4.11) confirms that GeOx-Ge quantum dots have adequate energy level separation and desired barrier height and width (of GeOx cladding) to permit charge transfer from channel to dots and among dots (via resonant tunneling, necessary for the manifestation of intermediate "i" state).

4.4.3 Thin Layer of SiN on Top of SiO_x-Cladded Si Dots on Top of SiO_2 Gate Insulator

The QDGFET works based on the direct tunneling of charge carriers from the inversion channel of the FET to the different layers of QDs and then to the gate terminal. The control gate insulator in this improved structure is too thin (20 Å) which allows the charge carriers to tunnel to the gate electrode. The basic functional difference between this improved structure of QDGFET and the quantum dot gate nonvolatile memory (QDNVM) is the thickness of control gate insulator. In the QDNVM, the control gate insulator is thicker than that of this new QDGFET. Because of this thickness difference, charge carriers cannot tunnel to the gate electrode of the QDNVM, whereas they can tunnel in the gate electrode of the QDGFET.

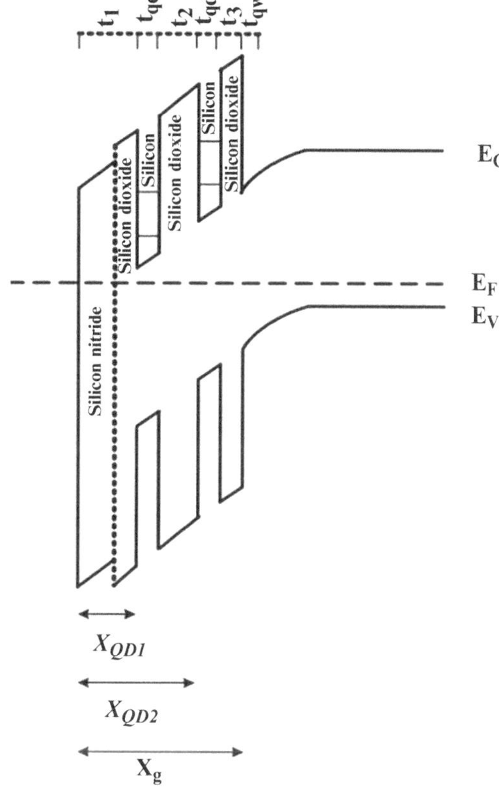

Fig. 4.13 Energy band diagram of the QDGFET, having silicon nitride on top of quantum dots in the gate region [18]

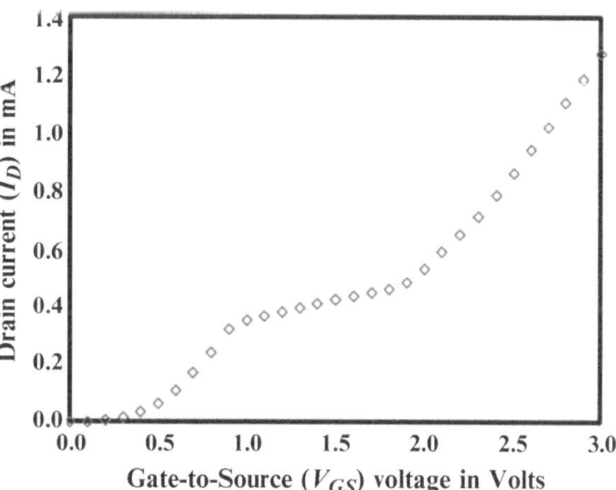

Fig. 4.14 Device model simulation results of the improved QDGFET structure [18]

The energy band diagram of the QDGFET, having silicon nitride on top of quantum dots in the gate region, is shown in Fig. 4.13. Figure 4.14 shows the device simulation result of this QDGFET.

References

1. Cobbold, R.S.C.: Theory and Applications of Field-Effect Transistors. Wiley-Interscience, New-York (1970). MOSFET band diagram and theory of operation
2. Kahng, D.: A historical perspective on the development of MOS transistors and related devices. IEEE Trans. Electron Devices **23**(7), 655–657 (1976)
3. Sah, C.T.: Characteristics of the metal-oxide-semiconductor transistor. IEEE Trans. Electron Devices **11**(7), 324–345 (1964)
4. Sah, Chih-Tang: A history of MOS transistor compact modeling. In: 2005 Nanotechnology Conference, Presentation Slides.
5. Frohman-Bentchkowsky, D., Vadasz, L.: Computer-aided design and characterization of digital MOS integrated circuits. IEEE J. Solid-State Circuits **4**(2), 57–64 (1969)
6. Jain, F., Papadimitrakopoulos, F.: Site-specific nanoparticle self-assembly. US Patent 7,368,370, 2008
7. Jain, F.C., Heller, E., Karmakar, S., Chandy, J.: Device and circuit modeling using novel 3-state quantum dot gate FETs. In: International Semiconductor Device Research Symposium, College Park, 12–15 Dec 2007
8. Chuang, S., Holonyak, N.: Efficient quantum well to quantum dot tunneling: analytical solutions. Appl. Phys. Lett. **80**, 1270–1272 (2002)
9. Hasaneen, E.-S., Heller, E., Bansal, R., Jain, F.: Modeling of nonvolatile floating gate quantum dot memory. Solid State Electron. **48**, 2055–2059 (2004)
10. Maserjian, J., Petersson, G.: Tunneling through thin MOS structures: dependence on energy. Appl. Phys. Lett. **25**, 50–52 (1974)
11. Grabert, H., et al.: Single Charge Tunneling: Coulomb Blockade Phenomena in Nanostructures. Series B: Physics, NATO ASI Series, vol. 294. Plenum Press, New York (1992)
12. Maserjian, J.: Tunneling in thin MOS structures. J. Vac. Sci. Technol. **11**(6), 996–1003 (1974)
13. Sune, J., Olivo, P., Ricco, B.: Quantum-mechanical modeling of accumulation layers in MOS structure. IEEE Trans. Electron Devices **39**, 1732–1739 (1992)
14. Weinberg, Z.A.: On tunneling in metal-oxide silicon structures. J. Appl. Phys. **53**(7), 5052–5056 (1982)
15. Register, L.F., et al.: Analytic model for direct tunneling in polycrystalline silicon-gate metal-oxide semiconductor devices. Appl. Phys. Lett. **74**(3), 457–459 (1999)
16. Lenzlinger, M., Snow, E.H.: Fowler–Nordheim tunneling into thermally grown SiO_2. J. Appl. Phys. **40**(1), 278–283 (1969)
17. Karmakar, S., Suarez, E., Jain, F.: Quantum dot gate three state FETs using ZnS – ZnMgS lattice-matched gate insulator on silicon. J. Electron. Mater. **40**(8), 1749–1756 (2011)
18. Karmakar, S., Gogna, M., Jain, F.C.: Improved device structure of quantum dot gate FET to get more stable intermediate state. Electron. Lett. **48**(24), 1556–1557 (2012)

Chapter 5
Quantum Dot Gate NMOS Inverter

This chapter introduces the NMOS inverter circuit based on QDGFETs. The three-state behavior of QDGFET also generates three states in the NMOS inverter based on this FET. The three-state nature of QDGFET-based NMOS inverter will help to implement multivalued logic in future. This chapter discusses the detailed fabrication methods of NMOS inverter.

5.1 Introduction

The inverter is the basic building block of any logic circuit design. An inverter [1, 2] based on conventional FET produces two states in their output characteristics, and they are useful for binary logic implementation.

The transfer characteristic of the QDGFET [3–5] shows three stable states ("0," "1," and "i"), where the low-current saturation state, "i," is manifested over a range of gate voltages which can be utilized for various circuit applications. The logic circuits using QDGFETs, such as an inverter [6–8], also show three states in their input-output characteristic. QDGFET-based logic circuits are compatible with CMOS processing technology. They can be used in the same conventional CMOS architecture to handle more number of bits.

This chapter presents a three-state logic inverter using two n-channel QDGFETs (QDNMOS inverter). The generation of the third state in the transfer characteristic of an NMOS inverter can be explained by the behavior of the QDGFET.

5.2 Conventional NMOS Inverter

Figure 5.1 shows the circuit diagram of a conventional NMOS inverter. In this circuit, the upper transistor acts as a load resistor which remains always on because of the high gate voltage. The lower transistor acts as a switch based on the applied

S. Karmakar, *Novel Three-state Quantum Dot Gate Field Effect Transistor: Fabrication, Modeling and Applications*, DOI 10.1007/978-81-322-1635-3_5,

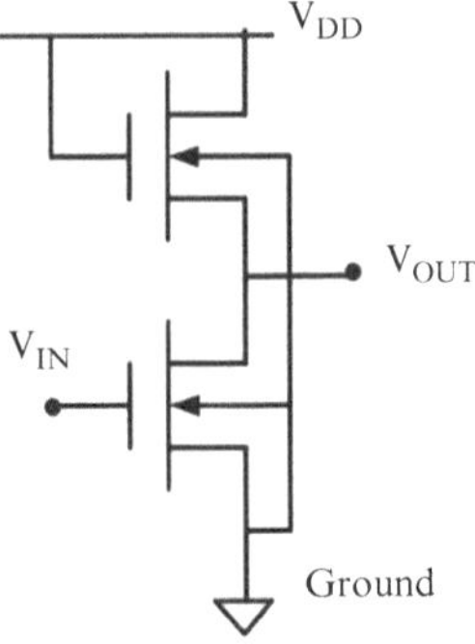

Fig. 5.1 Circuit diagram of a conventional NMOS inverter

input voltage. When the input voltage is higher than the threshold voltage of the lower transistor, the lower transistor turns on, connects the output to ground, and produces a low voltage at the output. When the input voltage is lower than the threshold voltage of the lower transistor, this transistor is off and output is high through the load (top) transistor.

5.3 QDNMOS Inverter

5.3.1 Device Structure

Figure 5.2 shows the cross-sectional schematic of the QDNMOS inverter [9]. Here, an n-type source and drain are used in a p-silicon (100) substrate. The gate consists of a thin layer of tunneling oxide followed by two layers of SiO_x-Si-cladded quantum dots [10–14].

Figure 5.3 shows the topology of the fabricated device. Here, A is the top transistor where the drain and the gate are connected by metal contact. The output terminal, V_{OUT}, is the source for the top transistor as well as the drain for the bottom transistor (B). The input terminal, V_{IN}, is the gate contact of bottom transistor (B), and the source of the bottom transistor (B) is connected to the ground.

5.3.2 Experimental Details for High-Resolution Transmission Electron Microscopy (HRTEM)

The high-resolution transmission electron microscope (HRTEM) image is obtained in the same way as discussed in Chap. 3.

The TEM image (Fig. 5.4) shows the deposition of two layers of SiO_x-cladded Si quantum dots in the gate region of the quantum dot gate FET.

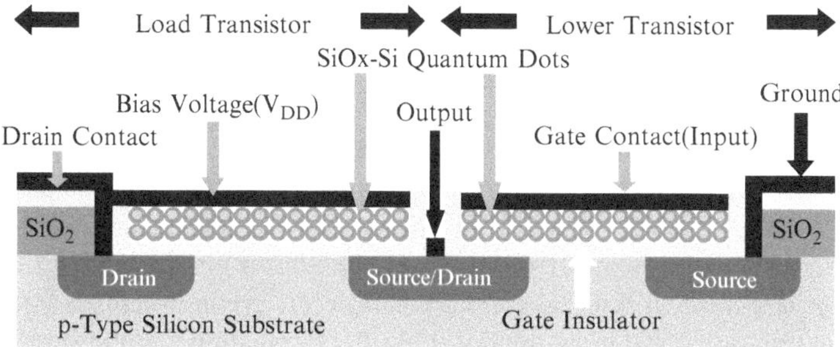

Fig. 5.2 Cross-sectional schematic of QDNMOS inverter [9]

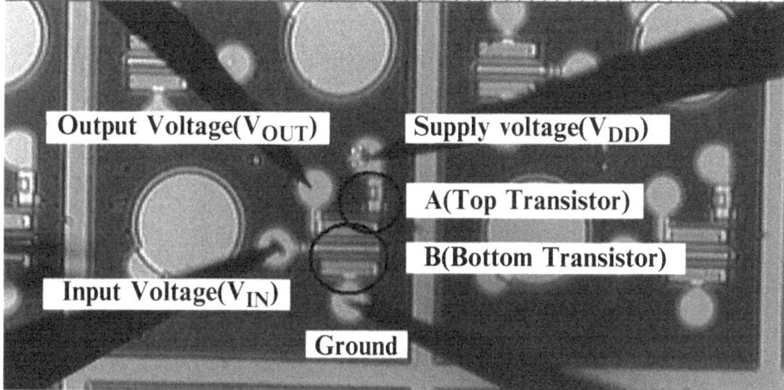

Fig. 5.3 Top view of fabricated QDNMOS inverter [9]

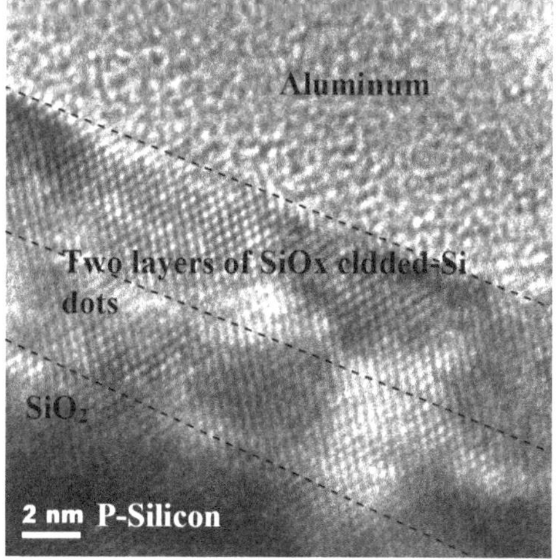

Fig. 5.4 HRTEM image of the gate region of the pull-down transistor

5.3.3 Fabrication Techniques

Phosphorous was diffused at 1,000 °C in a nitrogen environment to produce the n-type source and drain regions in a *p*-silicon (100) substrate. 20–30 Å tunneling oxide was grown on the gate region by dry oxidation at 800 °C. This process was followed by SiO_x-cladded Si quantum dot site-specific self-assembly. Aluminum was evaporated to form metal contacts for the source, drain, and gate regions. Annealing was done at 550 °C for 1 min in an N_2 environment to form ohmic contacts in the source and drain regions. Detailed fabrication methods are discussed as follows.

NMOS inverter based on quantum dot gate FET (QDGFET) is fabricated using four masks. Mask 1 is used to open source and drain for n-type doping. Mask 2 is used to open the gate window between source and drain. Mask 3 is used to open the source and drain contact window, and mask 4 is used to define different contact pads.

5.3.3.1 Cleaning of Silicon Wafer

A piece of p-type (100) silicon wafer (resistivity 10 Ω-cm) was used as substrate for this device. The cleaning process is same as silicon wafer cleaning discussed in Chap. 3.

5.3.3.2 Source and Drain Processing

1. Wet oxide was grown at 1,000 °C for 15 min which grew 170-Å oxide on top of the sample (Fig. 5.5B). This wet oxide acts as a mask for phosphorous diffusion in the source-drain region and also provides good visibility of different patterns for the following photolithography process.
2. A 2-in. mounting wafer was placed on a spinner and spun with S1813 positive photoresist at 1,000 rpm for 10 s.
3. The sample was then placed in the center of the mounting wafer and placed on hot plate at 115 °C for 5–6 min until the sample adhered to the wafer.
4. The mounting wafer was then placed on the spinner and spin-coated the sample with S1813 positive photoresist at 5,000 rpm for 30 s (Fig. 5.5C).
5. The mounting wafer with the sample was then baked (preexpose bake) at 115 °C for 2 min.
6. Exposed the sample in UV for 30 s with mask 1 (Fig. 5.5D).
7. Pre-develop baked at 115 °C for 1 min.
8. Developed the sample in 3.5:1 (water to 351 Developer) developer solution for 10–15 s (Fig. 5.5E).
9. Rinsed thoroughly in DI water to clean developer solution and post-baked the sample for 10 min at 115 °C.

10. Etched field oxide grown by wet oxidation process from source and drain region using buffered oxide etch for a few minutes and check whether the oxide had been removed (Fig. 5.5F).
11. Removed sample from mounting wafer using acetone.
12. Cleaned the photoresist using acetone. Then clean with methanol and DI water.
13. Placed the sample in boiling propanol for transferring it to phosphorous diffusion furnace.
14. The sample was dried off and placed on a boat with phosphorous source. The sample was then loaded in phosphorous diffusion furnace. Phosphorous diffusion was done at 1,000 °C for 5 min in a nitrogen (N_2) environment (Fig. 5.5G).

5.3.3.3 Gate Processing

The source-drain diffusion is followed by gate processing.

1. Steps 2–3 of the source-drain processing were performed using mask 2 (gate opening mask).
2. The mounting wafer was then placed on a spinner and spin-coated the sample with AZ5214 negative photoresist at 5,000 rpm for 20 s.
3. The mounting wafer with the sample was then baked (preexpose bake) at 90 °C for 4 min.
4. Exposed the sample in UV for 25 s with mask 2 (Fig. 5.5H).
5. Pre-develop baked at 115 °C for 2 min.
6. Flood-exposed the sample under the UV having power 2,000–2,400 μW/cm^2 and wavelength 365 nm for 75 s.

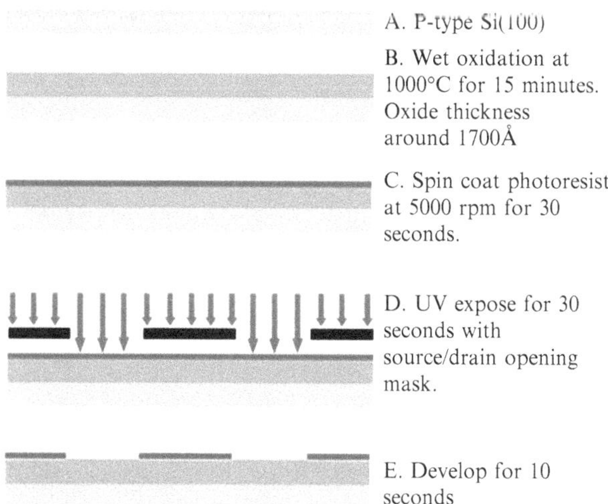

Fig. 5.5 Process flow for NMOS inverter fabrication

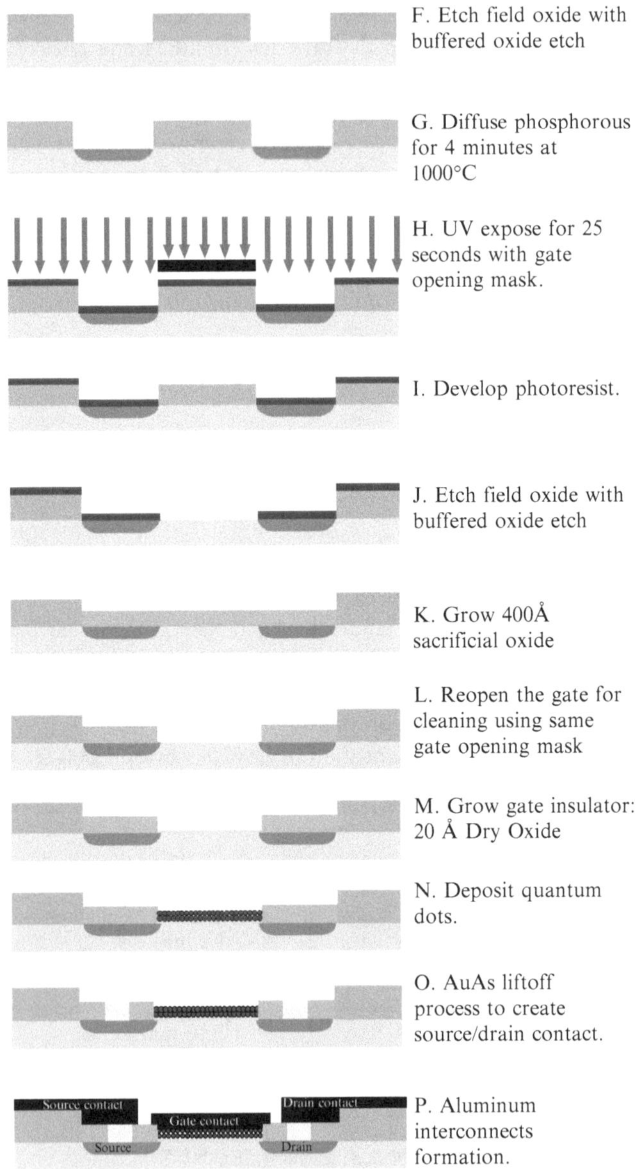

Fig. 5.5 (continued)

7. Developed the sample in 1:1 (water to AZ Developer) developer solution for 30 s–1 min (Fig. 5.5I).
8. Rinsed thoroughly in DI water to clean developer solution and post-baked the sample for 4 min at 115 °C.

9. Etched field oxide grown by wet oxidation process from gate region using buffered oxide etch for few minutes and check whether the oxide had been removed (Fig. 5.5J).
10. Removed sample from mounting wafer using acetone.
11. Cleaned the photoresist using acetone. Then cleaned with methanol and DI water.
12. Sample was stored in boiling isopropyl alcohol and transferred to dry oxidation furnace for sacrificial gate oxide growth.
13. The sample was then placed into oxidation furnace in oxygen environment at 900 °C for 20 min and 1,050 °C for 10 min. This oxidation process grew approximately 400 Å gate oxide (Fig. 5.5K).

Reopening Gate Region

The sacrificial oxide grown in the previous step was to clean the gate region. The growth of sacrificial oxide was followed by reopening the gate region and grew tunnel insulator (Fig. 5.5L).

1. Steps 2–12 of this process were performed.
2. The sample was then placed into oxidation furnace in oxygen environment at 800 °C for 5 min. This oxidation process grew approximately 20 Å gate oxide (Fig. 5.5M).

SiOx-Cladded Si Dot Self-Assembly

The SiO_X-cladded Si quantum dots were site-specifically self-assembled on top of tunnel oxide in the gate region. This process was same as discussed in section 3.1.1.3 in this thesis (Fig. 5.5N).

5.3.3.4 Source-Drain Contact Formation

Source and drain contact were formed by opening the source and drain contact hole by using source-drain contact hole opening mask and etching the deposited silicon dioxide and quantum dot layers on top of the source and drain region. Gold arsenic was thermally evaporated on top of sample, and source-drain contact deposition was done by liftoff method. The final ohmic contact was formed by annealing the contacts at 375 °C for 1 min in nitrogen-hydrogen (N_2:H_2 = 9:1) environment (Fig. 5.5O).

Finally, aluminum evaporation was done to form the gate contact and source-drain contact pads (Fig. 5.5P).

Figure 5.6 shows the circuit diagram of the NMOS inverter based on QDGFET. Figure 5.7 shows the measured transfer characteristics of the fabricated

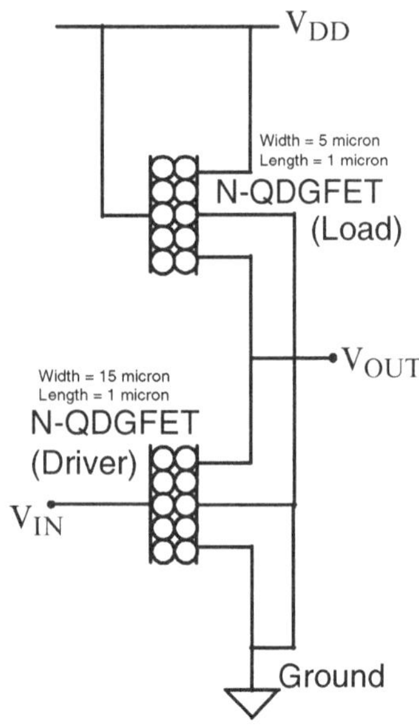

Fig. 5.6 Circuit diagram of QDNMOS inverter

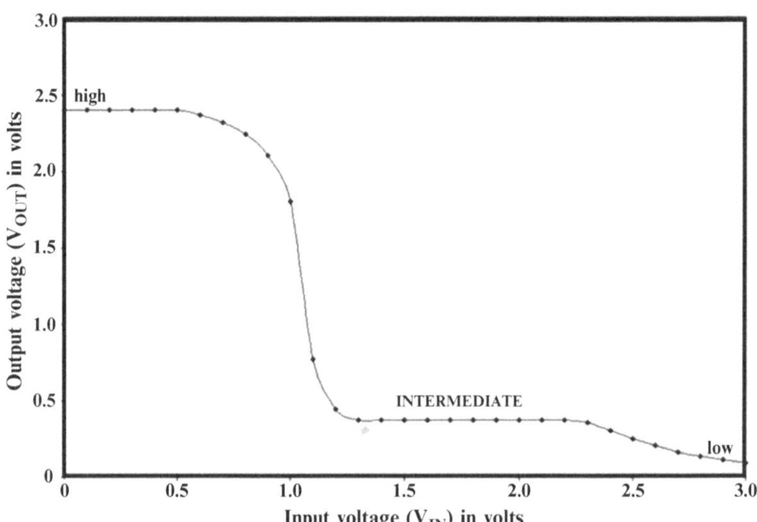

Fig. 5.7 Three states in the output characteristic of NMOS inverter based on QDGFET [9]

QDGFET-based NMOS inverter. The output voltages are 2.4, 0.378, and 0.095 V when the input voltage is in low, intermediate, and high voltage range, respectively. Optimization is needed to adjust the spacing of on, intermediate, and off states.

References

1. Kruppa, W., Boos, J. B.: Observation of DC and microwave negative differential resistance in InAlAs/InGaAS/InP HEMTs. Electron. Lett. **38**(3), 267–269 (1992)
2. Chumbes, E.M., Schremer, A.T., Smart, J.A., Wang, Y., MacDonald, N.C., Hogue, D., Komiak, J.J., Lichwalla, S.J., Leoni, R.E., Shealy, J.R.: AlGaN/GaN high electron mobility transistors on Si(111) substrates. IEEE Trans. Electron Devices **48**(3), 420–425 (2001)
3. Jain, F.C., Suarez, E., Gogna, M., AlAmoody, F., Butkiewicus, D., Hohner, R., Liaskas, T., Karmakar, S., Chan, P.Y., Miller, B., Chandy, J., Heller, E.: Novel quantum dot gate FETs and nonvolatile memories using lattice-matched II–VI gate insulators. J. Electron. Mater. **38**(8), 1574–1578 (2009)
4. Karmakar, S., Gogna, M., Suarez, E., Alamoody, F., Heller, E., Chandy, J., Jain, F.: 3-state behavior of quantum dot gate FETs with lattice matched insulator. In: Proceedings of 2009 Nanoelectronic Devices for Defense and Security, Fort Lauderdale, Florida, USA, Sept 2009
5. Jain, F., Karmakar, S., Alamoody, F., Suarez, E., Gogna, M., Chan, P.-Y., Chandy, J., Miller, B., Heller, E.: 3-state behavior in quantum dot gate InGaAs FETs. In: Proceedings of International Semiconductor Device Research Symposium, College Park, MD, USA, Dec 2009
6. Jain, F.C., Heller, E., Karmakar, S., Chandy, J.: Device and circuit modeling using novel 3-state quantum DOT gate FETs. In Proceedings of International Semiconductor Device Research Symposium, College Park, MD, USA, Dec 2007
7. Chandy, J.A., Jain, F.C.: Multiple valued logic using 3-state quantum dot gate FETs. In: Proceedings of International Symposium on Multiple Valued Logic, Dallas, Texas, USA, 08
8. Karmakar, S., Suresh, A.P., Chandy, J.A., Jain, F.C.: Design of ADCs and DACs using 3-state quantum DOT gate FETs. In: Proceedings of International Semiconductor Device Research Symposium, College Park, MD, USA, Dec 2009
9. Karmakar, S., Chandy, J.A., Gogna, M., Jain, F.C.: Fabrication and Circuit modeling of NMOS inverter based on quantum dot gate field effect transistors. J. Electron. Mater. **41**(8), 2184–2192 (2012). doi:10.1007/s11664-012-2116-4
10. Gogna, M., Al-Amoody, F., Karmakar, S., Papadimitrakopoulos, F., Jain, F.: Quantum Dot (QD) gate Si-FETs with self-assembled GeO_X Cladded Germanium Quantum Dots. In: Nanotech 2009 Conference, Houston, TX, USA, **1**, 163–165
11. Jain, F., Alamoody, F., Suarez, E., Gogna, M., Chan, P.-Y., Karmakar, S., Fikiet, J., Miller, B., Heller, E.: Quantum dot gate InGaAs FETs. In: *Nanotech 2009 Conference*, Houston, TX, USA, **1**, 598–601
12. Papadimitrakopoulos, F., Phely-Bobin, T., Wisniecki, P.: Self-assembled nanosilicon/siloxane composite films. Chem. Mater. **11**(3), 522–525 (1999)
13. Phely-Bobin, T., Chattopadhyay, D., Papadimitrakopoulos, F.: Characterization of mechanically attrited Si/SiOx nanoparticles and their self-assembled composite films. Chem. Mater. **14**(3), 1030–1036 (2002)
14. Jain, F.C., Papadimitrakopoulos, F.: US Patent 7,368,370, 2008

Chapter 6
Quantum Dot Gate Field-Effect Transistor (QDGFET): Circuit Model and Ternary Logic Inverter

The circuit model of QDGFETs is introduced in this chapter. This chapter highlights the modification of BSIM MOSFET model based on the characteristic of the QDGFET. Standard ternary logic inverter (STI) is also discussed in this chapter. The circuit simulation of NMOS inverter and the comparison between experimental data and simulation results is also presented in this chapter. The drawback of QDGFET-based STI is also discussed in this chapter which is followed by the three-state memory cell based on QDGFET.

BSIM [1, 2] is a physics-based, accurate, scalable, robustic, and predictive MOSFET SPICE model for circuit simulation and CMOS technology development. It is developed by the BSIM Research Group in the Department of Electrical Engineering and Computer Sciences (EECS) at the University of California, Berkeley. The third iteration of BSIM3, BSIM3 Version 3 (commonly abbreviated as BSIM3v3), was established by SEMATECH as the first industry-wide standard of its kind in December of 1996. BSIM3v3 has since been widely used by most semiconductor and IC design companies worldwide for device modeling and CMOS IC design. This chapter provides the modification of BSIM model to simulate the QDGFET electrical characteristics.

6.1 QDGFET Circuit Model

An empirical circuit model [3, 4] is developed for the QDGFET that accounts for the intermediate state "i" which manifests within the range of gate voltages V_{g1} and V_{g2}. The effective threshold voltage is divided into three ranges corresponding to the three regions of the transfer characteristics (see Fig. 6.3): region 1, intermediate state "i," and the saturation region.

S. Karmakar, *Novel Three-state Quantum Dot Gate Field Effect Transistor: Fabrication, Modeling and Applications*, DOI 10.1007/978-81-322-1635-3_6,

$$V_{Teff} = \begin{cases} V_T & V_{GS} < V_{g1} \\ V_T + \alpha\left(V_{GS} - V_{g1}\right) & V_{g1} < V_{GS} < V_{g2} \\ V_T + \alpha\left(V_{g2} - V_{g1}\right) & V_{GS} > V_{g2} \end{cases} \tag{6.1}$$

As the gate voltage increases through a range of voltages (V_{g1} to V_{g2}), the threshold voltage changes linearly with respect to the gate voltage. The effect is controlled by the α parameter. With $\alpha = 0$, the QDGFET behaves like a conventional FET, and with $\alpha = 1$, the QDGFET has a threshold voltage that changes directly with the gate voltage. This α parameter can be controlled by the thickness of various insulators, and by changing the size and number of dots, that results in changing of the QD gate charge. Likewise, the V_{g1} and V_{g2} values are also determined by the device structure. Using the equations for V_{Teff} from Eq. 6.2, we can give the drain current equations using traditional MOSFET circuit modeling techniques.

$$I_{DS} = \begin{cases} 0 & V_{GS} < V_{Teff} \\ \frac{W}{L} C_o \mu \left(V_{GS} - V_{Teff} - \frac{V_{DS}}{2} \right) V_{DS} & V_{DS} < V_{GS} - V_{Teff} \\ \frac{W}{L} C_o \mu \frac{\left(V_{GS} - V_{Teff}\right)^2}{2} & V_{DS} > V_{GS} - V_{Teff} \end{cases} \tag{6.2}$$

When $\alpha = 1$, $V_{Teff} = V_T + V_{GS} - V_{g1}$, meaning that I_{DS} $\left(I_{DS} = \frac{W}{L} C_o \mu \left(V_{g1} - V_T - \frac{V_{DS}}{2}\right) V_{DS}\right)$ does not depend on V_{GS} over the V_{g1} to V_{g2} region.

As can be seen, I_{DS} is not dependent on V_{GS} over this region, leading to a flat current curve over the region. Using these principles, we have constructed a Verilog AHDL (analog hardware description language) QDGFET model and conducted simulations of QDGFET-based complement functions and memory cells. The first step in utilizing these QDGFETs in a multivalued logic circuit is to determine the voltage levels associated with the multivalued logics. In the designs that we have presented below, we assume that logic level 0 corresponds to 0 V or ground, logic level 1 corresponds to $V_{DD}/2$ (the intermediate state i), and finally logic level 2 corresponds to V_{DD}.

In this research work, an improved circuit model for QDGFET was developed considering all effects based on the Berkeley simulation model (BSIM 3.2.0 and BSIM 3.2.4) for nano-FET modeling, and simulations of different ternary logic circuits based on QDGFET are conducted in Cadence. Table 6.1 shows different parameters for circuit model simulation of QDGFET. Figure 6.1 shows the comparison of the circuit model with the fabricated device transfer characteristics.

Table 6.1 Different parameters for QDGFET

	N-QDGFET	P-QDGFET
Mobility	600	300
Minimum L	5 μm	5 μm
Minimum W	5 μm	5 μm
V_T	0.2 V	0.2 V
V_{g1}	0.6 V	0.6 V
V_{g2}	1.2 V	1.2 V
V_{DD}	5 V	5 V
α	1.0	1.0

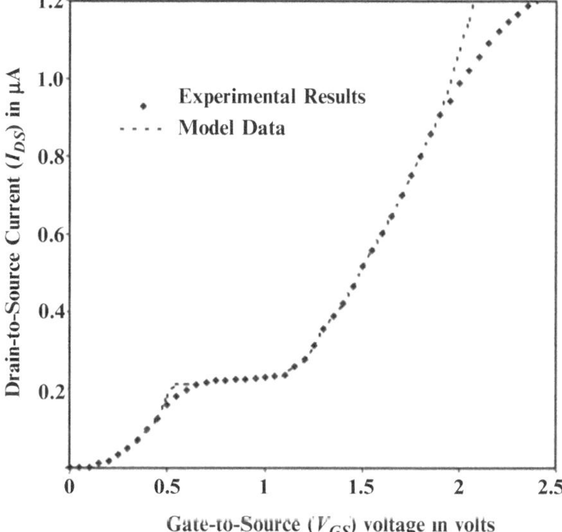

Fig. 6.1 Comparison of quantum dot gate FET (QDGFET) circuit model with fabricated device. --- model data and ■ ■ ■ ■ fabricated device characteristic

6.2 Inverter

The inverter is truly the nucleus of all digital designs [5, 6]. Once its operation and properties are clearly understood, designing more intricate structures such as logic gates, adders, multipliers, and microprocessors is greatly simplified. The electrical behavior of these complex circuits can be almost completely derived by extrapolating the results obtained for inverters. The analysis of inverters can be extended to explain the behavior of more complex gates such as NAND, NOR, and XOR, which in turn form the building blocks for modules such as multipliers and processors.

In this chapter, the ternary logic static CMOS inverter is discussed in detail. Before discussing ternary logic inverter, static CMOS inverter will be discussed to some extent.

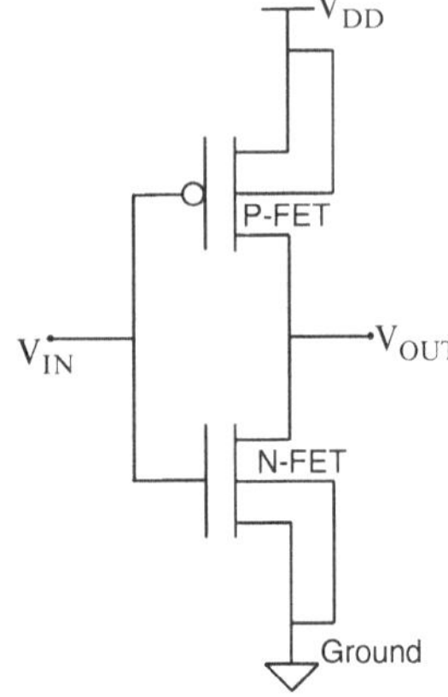

Fig. 6.2 CMOS inverter circuit where V_{DD} is the supply voltage

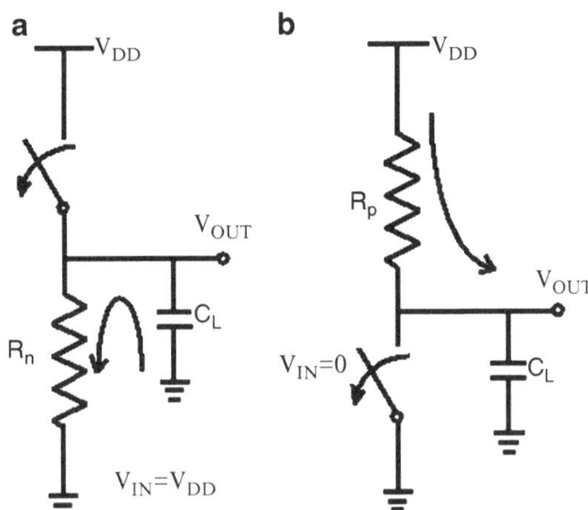

Fig. 6.3 Switch model of dynamic behavior of CMOS inverter. (**a**) Output goes from high to low voltage. (**b**) Output goes from low to high voltage

6.2.1 *The Static CMOS Inverter*

The circuit diagram [6] of a static CMOS inverter is shown in Fig. 6.2. The operation of a static CMOS inverter can be explained by the switching operation of MOS transistor. A MOS transistor has an infinite off resistance when gate-to-source voltage is greater than the threshold voltage of the transistor ($|V_{GS}| < |V_T|$) and finite on resistance when gate-to-source voltage is greater than the threshold voltage of the transistor ($|V_{GS}| > |V_T|$). When the input voltage at the gate terminal is high (equals to V_{DD}), the NMOS is on and the PMOS is off. The equivalent circuit diagram is shown in Fig. 6.3a. In this condition, a direct path exists between the V_{OUT} and the ground node via the NMOS transistor, and the output becomes 0 V. On the other hand, when the input voltage is low (0 V), the NMOS transistor is off and the PMOS transistor is on. The equivalent circuit diagram is shown in Fig. 6.3b which shows the existence of a path between V_{DD} and V_{OUT}, which makes the output high.

The basic advantage of a complementary CMOS gate is the rail-to-rail swing of its different voltage levels which increases its noise margin in different logic operation. The circuit is also free from any static power because the circuits are designed in such a way that the pull-up and the pull-down networks are mutually exclusive.

Although the CMOS architecture is very simple and robust, there are two major disadvantages of complementary CMOS architecture. One is the number of gates. The number of gates is always twice (2N) than the number of fan-in (N) which results in a significantly large implementation area. Other major problem is propagation delay. The propagation delay is a quadratic function of the number of fan-in (N).

According to this architecture, for an N input gate, the number of transistor should be 2N. The intrinsic capacitance increases linearly with the number of transistor. As the number of CMOS connected to the output node increases linearly, the low-to-high delay also increases linearly.

In different network architecture, the series connection of transistors causes additional delay. The delay is a quadratic function of the number of elements in the network.

6.2.2 Ratioed Logic: NMOS Inverter

The number of transistors can be reduced by implementing ratioed logic in digital logic circuit design. This costs reduced robustness and extra power dissipation of the designed circuit. In the complementary CMOS architecture, pull-up network provides a path between V_{DD} and output when the transistor is off.

In ratioed logic, the PUN is replaced by a load device that pulls up the output for a high output voltage. In the logic design, the logic function is implemented by the pull-down network, and a simple load device replaces the pull-up network in CMOS architecture.

The advantage of pseudo-NMOS gate architecture is that the number of circuit elements is reduced over CMOS architecture. The main disadvantage of pseudo-NMOS architecture is the reduced robustness. The high output voltage in this architecture is V_{DD}, whereas the low output voltage is not 0 V. As a result the noise margin for this architecture is less than that of CMOS architecture. Besides this, the static current flow in this circuit when PDN is on is a major concern.

Different performance parameters such as noise margin, propagation delay, and power dissipation can be controlled by the sizing of the load device relative to the pull-down devices. Since in this circuit architecture the overall functionality and the output swing of the gate depend on the sizing ratio of the PMOS and NMOS, this logic is known as ratioed logic.

A larger pull-up device not only improves performance but also increases static power dissipation and lowers noise margins. When area is most important, however, its reduced transistor count compared to complementary CMOS is quite attractive.

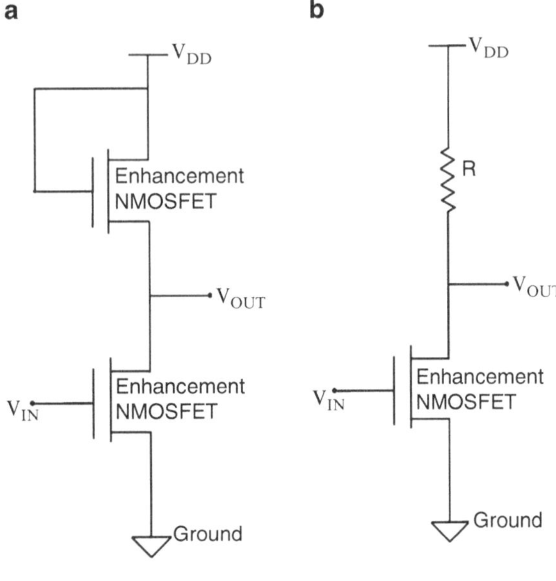

Fig. 6.4 (**a**) Pseudo-NMOS gate NMOS inverter, (**b**) the load is replaced by a simple resistor

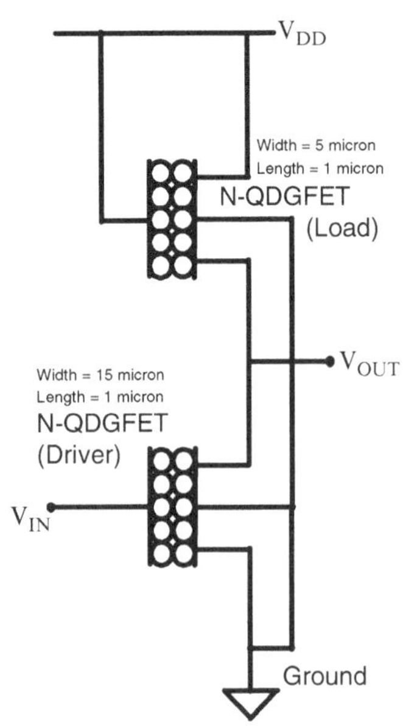

Fig. 6.5 Circuit diagram of QDNMOS inverter

In simulations, the conventional NMOS FET in an NMOS inverter circuit (Fig. 6.4) is replaced with a QDGFET characterized by a BSIM model. Figure 6.5 shows the NMOS inverter based on quantum dot gate FET (QDGFET). The circuit was simulated using the Cadence Analog Environment. Table 6.2 shows the different parameters for simulation of QDGFET-based NMOS inverter.

Table 6.2 NMOS inverter simulation parameters

	Top transistor	Bottom transistor
V_{TH} (threshold voltage)	0.6 V	0.6 V
V_{g1} (lower threshold voltage)	1.3 V	1.3 V
V_{g2} (upper threshold voltage)	2.3 V	2.3 V
T_{ox} (gate insulator thickness)	20 Å	20 Å
μ (cm^2/Vs) (mobility)	600	600
W (width)	5 um	15 um
L (length)	1 um	1 um

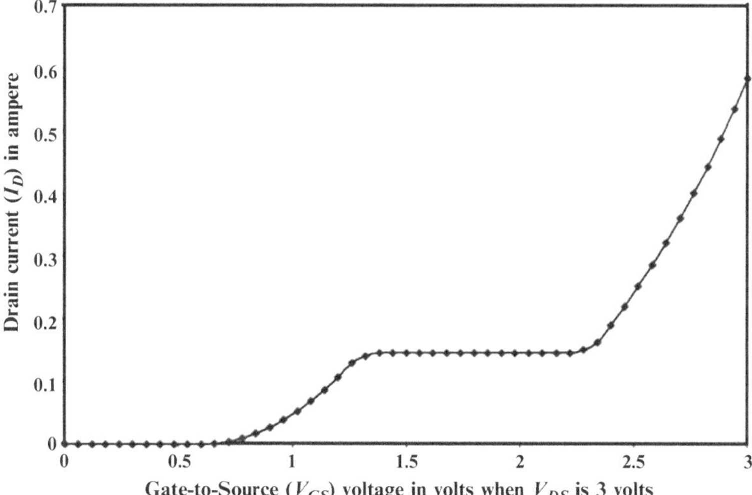

Fig. 6.6 Transfer characteristic ($I_D - V_{GS}$) of the load transistor (*top* QDGFET)

6.2.2.1 Operation Principle

In the fabricated device (Chap. 5), the load transistor (top transistor) width/length ratio is one third of the width/length ratio of the driver transistor (bottom transistor). Both transistors produce three states in their transfer ($I_D - V_{GS}$) characteristics because of the presence of two layers of SiO_x-cladded Si quantum dots in the gate region.

Figures 6.6 and 6.7 show the transfer characteristics of the load (top) and the driver (bottom) transistors respectively in the fabricated device. As can be seen, the current in both transistors has an intermediate state as predicted. Figures 6.8 and 6.9 show the output characteristics of the top and bottom transistors respectively. Unlike a normal FET, we can see the curves are clustered together when V_{GS} is in the intermediate region. Both transistors have 0.6 V threshold voltage.

When the input voltage (V_{IN}) of the inverter is low, the driver transistor (lower) is off, so the output is pulled up by the top transistor to ($V_{DD} - V_{TH}$). The threshold voltage for the top transistor is ~0.6 V, so the output will be 2.4 V when the input voltage is low.

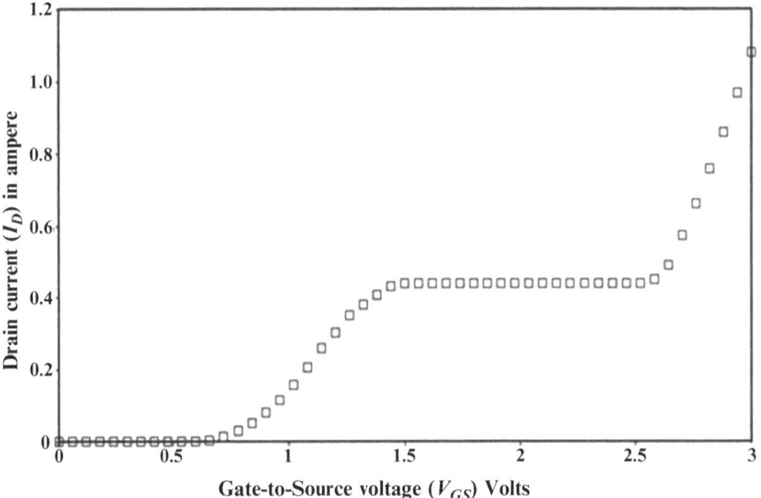

Fig. 6.7 Transfer characteristic ($I_D - V_{GS}$) of the driver (*bottom* QDGFET) when $V_{DS} = 3$ V

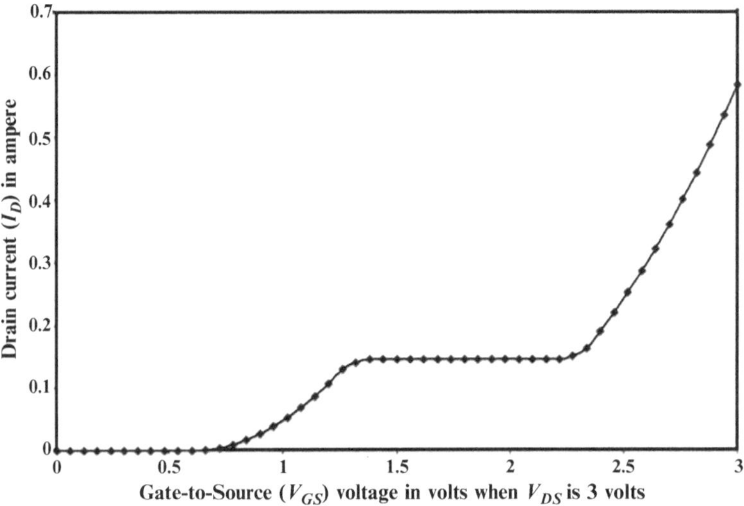

Fig. 6.8 Output characteristics ($I_D - V_{DS}$) of the load transistor (*top* QDGFET)

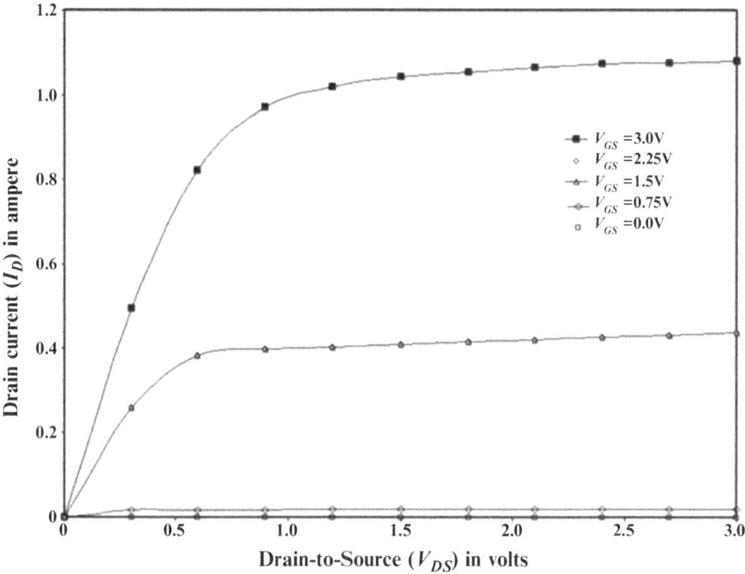

Fig. 6.9 Output characteristics ($I_D - V_{DS}$) of the driver transistor (*bottom* QDGFET)

When the input voltage is in the intermediate range or high, the load transistor (upper) is always in the saturation region. Drain-to-source current is expressed as

$$I_{DS} = \left(\frac{W}{L}\right)_{\text{Load}} C_{\text{OX}}\mu \frac{\left(V_{GS} - V_{\text{TeffLoad}}\right)^2}{2} \tag{6.3}$$

where

$$V_{GS} = V_{\text{DD}} - V_{\text{OUT}}. \tag{6.4}$$

Substituting the value of V_{GS} in Eq. 6.3, I_{DS} of the top transistor can be expressed as in Eq. 6.5:

$$I_{DS\,\text{Load}} = \beta_{\text{T}} C_{\text{OX}}\mu \frac{\left(V_{\text{DD}} - V_{\text{OUT}} - V_{\text{TeffLoad}}\right)^2}{2} \tag{6.5}$$

where

$$\beta_{\text{T}} = \left(\frac{W}{L}\right)_{\text{Load}} \tag{6.6}$$

In this input voltage range, the lower transistor should be in the linear region. The drain-to-source current for the bottom transistor can be expressed according to Eq. 6.7:

$$I_{DS\,\mathrm{Driver}} = \beta_{\mathrm{B}} C_{\mathrm{OX}} \mu \left(V_{\mathrm{IN}} - V_{\mathrm{TeffDriver}} - \frac{V_{\mathrm{OUT}}}{2} \right) V_{\mathrm{OUT}} \tag{6.7}$$

where

$$\beta_{\mathrm{B}} = \left(\frac{W}{L} \right)_{\mathrm{Driver}} \tag{6.8}$$

At the switching time, the same current will flow through both transistors. So,

$$I_{DS\,\mathrm{Load}} = I_{DS\,\mathrm{Driver}} \tag{6.9}$$

Equating Eqs. 6.5 and 6.7

$$\beta_{\mathrm{T}} C_{\mathrm{OX}} \mu \frac{(V_{\mathrm{DD}} - V_{\mathrm{OUT}} - V_{\mathrm{TeffLoad}})^2}{2} = \beta_{\mathrm{B}} C_{\mathrm{OX}} \mu \left(V_{\mathrm{IN}} - V_{\mathrm{TeffDriver}} - \frac{V_{\mathrm{OUT}}}{2} \right) V_{\mathrm{OUT}} \tag{6.10}$$

If we solve for V_{OUT} from Eq. 6.10,

$$V_{\mathrm{OUT}} = \frac{[\beta_{\mathrm{R}} V_{\mathrm{DL}} - V_{\mathrm{IB}}] \pm \sqrt{V_{\mathrm{IB}}^2 - 2\beta_{\mathrm{R}} V_{\mathrm{DL}} V_{\mathrm{IB}} - 2\beta_{\mathrm{R}} V_{\mathrm{DL}}^2}}{(2 + \beta_{\mathrm{R}})} \tag{6.11}$$

where

$$\beta_{\mathrm{R}} = \frac{\beta_{\mathrm{T}}}{\beta_{\mathrm{B}}} = \frac{1}{3}, \quad V_{\mathrm{DL}} = V_{\mathrm{DD}} - V_{\mathrm{TeffLoad}} \text{ and } V_{\mathrm{IB}} = V_{\mathrm{IN}} - V_{\mathrm{TeffDriver}}. \tag{6.12}$$

Figure 6.10 shows the comparison between the fabricated device transfer ($V_{OUT} - V_{IN}$) characteristic and the dc transfer characteristics of QDNMOS inverter, which shows a similar output.

According to the above computation, the output voltage should be ~0.39 V.

6.3 Ternary Inversion Operation

A ternary inversion is an operation with one input (r) and three outputs (l_0, l_1, l_2) such that

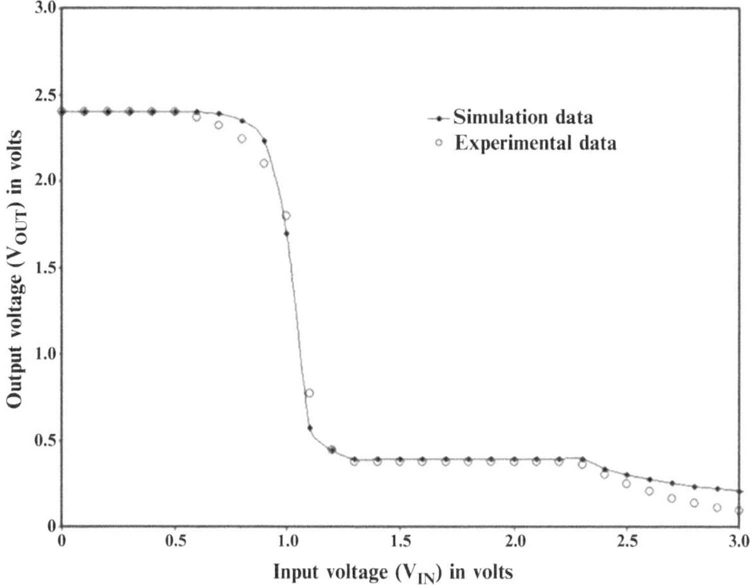

Fig. 6.10 Comparison between experimental data ($V_{OUT} - V_{IN}$) and the simulation results of QDGFET-based NMOS inverter

$$\begin{aligned} l_0 &= \begin{cases} 2 & if \quad r = 0 \\ 0 & if \quad r \neq 0 \end{cases} && \text{(a)} \\ l_1 &= 2 - r && \text{(b)} \\ l_2 &= \begin{cases} 2 & if \quad r \neq 2 \\ 0 & if \quad r = 2 \end{cases} && \text{(c)} \end{aligned} \tag{6.13}$$

where r is the number of states in the logic space [7]. The implementation of ternary inverters requires three conventional inverters: negative ternary inverter (NTI), standard ternary inverter (STI), and positive ternary inverter (PTI) where l_0, l_1, and l_2 should be the outputs, respectively [8]. The truth table for these three inverters is shown in Table 6.3.

The circuit diagram of a standard ternary inverter based on QDGFET is shown in Fig. 6.11. This circuit is the same as a conventional CMOS inverter, except that the transistors have been replaced by quantum dot gate FETs (QDGFET). In this circuit, when the input is "0," the P-QDGFET is on and the N-QDGFET is in the off state, which makes the output "2." When the input is "2," the P-QDGFET is on and the N-QDGFET is off, which makes the output "0." When V_{IN} equals to "1," both transistors are in the intermediate mode, thus making both of them behave like a resistor. In this situation, the circuit behaves like a voltage divider which produces

Table 6.3 Ternary logic inverter truth table

Input	STI	PTI	NTI
0	2	2	2
1	1	2	0
2	0	0	0

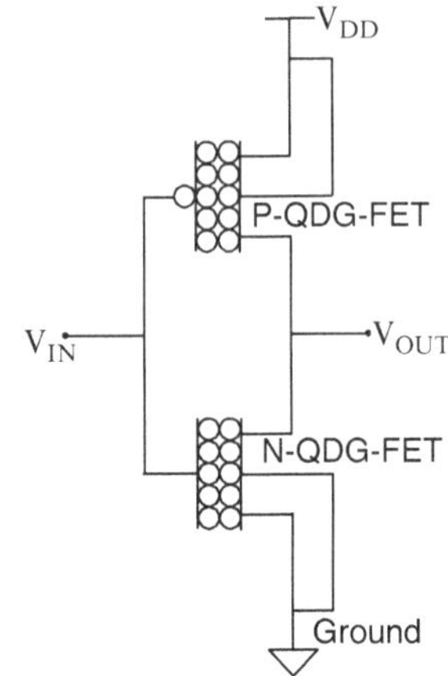

Fig. 6.11 CMOS architecture using quantum dot gate FET (QDGFET)

an intermediate "1" logic output, assuming both transistors have same β values. Over this range, ideally we would like to see a flat curve. However, to maintain stability, we only require that the slope of the curve is less than −1. If we establish $V_{DD}/2$ as the intermediate voltage point, this design will provide a noise margin of 0.5 V and still keep the output within the acceptable range. Because the slope is less than −1, a chain of complement functions will self-correct the voltage to $V_{DD}/2$. It is clear that the QDGFET complement function provides a stable intermediate logic output voltage and can be used as the basis for multivalued logic. The transfer characteristic of the complement function is shown in Fig. 6.12. The transistor parameters used in the simulation are shown in Table 6.1.

However, there are two drawbacks to the QDGFET complement function as compared to a CMOS inverter. The first is that QDGFET complement function loses the ratioless property of a CMOS inverter. In other words, changing the size ratio between the n and p transistors will alter the behavior of the complement function. Figure 6.13 shows the effect of changing the width of the p transistor from 1,540 to 900 nm. Reducing the width of the transistor effectively increases the resistance of the transistor with respect to the voltage divider, thus pushing the intermediate voltage down. Thus, one must always maintain the correct ratio between p and n transistors when making sizing decisions for performance or other reasons. This is a disadvantage with respect to CMOS designs, but is not significant, since in most current CMOS designs, it is rare to size the n and p transistors independently.

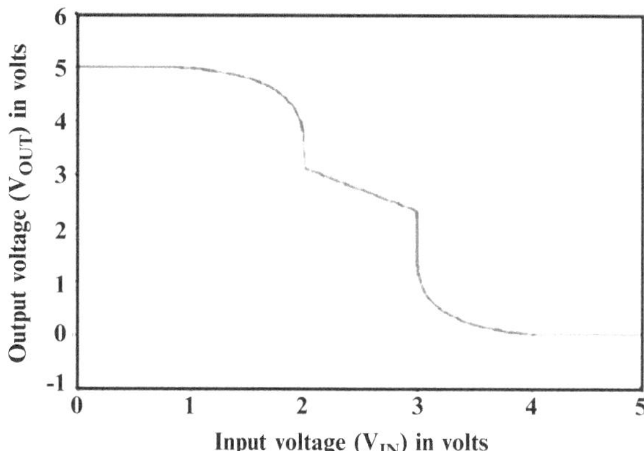

Fig. 6.12 Transfer characteristic of standard ternary inverter (STI) based on QDGFET

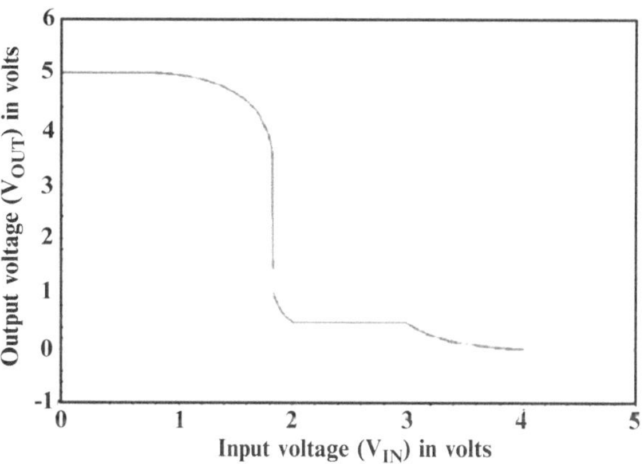

Fig. 6.13 Transfer characteristic of standard ternary inverter (STI) with smaller P-QDGFET

The other major limitation with QDGFET designs is static current draw when the output of the complement function is at logic level 1. Figure 6.14 shows the current for the complement function. At logic states "0" and "2," there is no static current flow, but at state "1," the intermediate state, there is static current. This current is less than what would be seen in a conventional NMOS inverter, because the quantum dot gate transistor does not enter saturation. Static current draw is a typical problem of many MVL circuit structures [9, 10].

6.4 Three-State Memory Cell

In this section, the use of QDGFETs in the design of multivalued logic state memory is presented. Particularly, we are interested in the behavior of back-to-back complement functions as a state memory (Fig. 6.15). Figure 6.16 shows the transfer characteristics of back-to-back complement functions. Over the

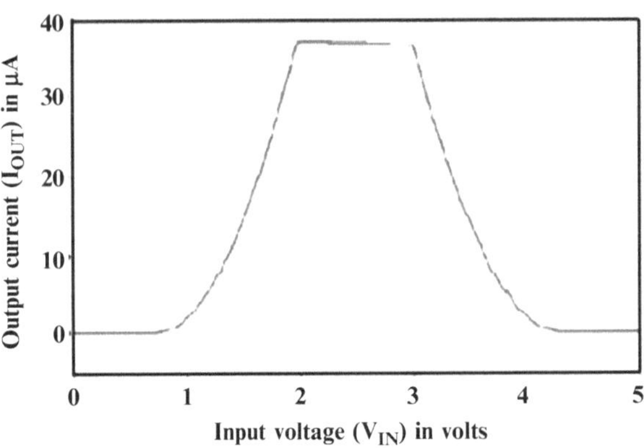

Fig. 6.14 Transfer characteristics showing drain leakage current in the intermediate state

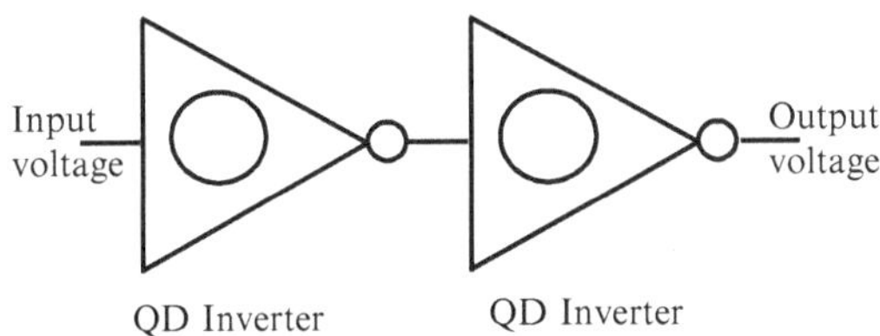

Fig. 6.15 Block diagram of back-to-back complement functions

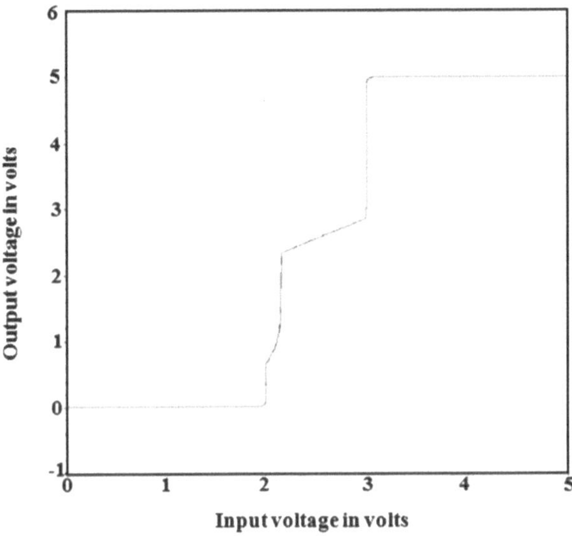

Fig. 6.16 Transfer characteristics of back-to-back complement function

intermediate input range between 2 and 3 V, the output ranges from 2.2 to 2.8 V which when fed back into the first complement function will reduce the range until the value settles around 2.5 V. Thus, the back-to-back complement functions will serve as an adequate three-state memory. The flat I_{DS} curve provides the relatively flat $V_{OUT}-V_{IN}$ curve that allows this feedback to work.

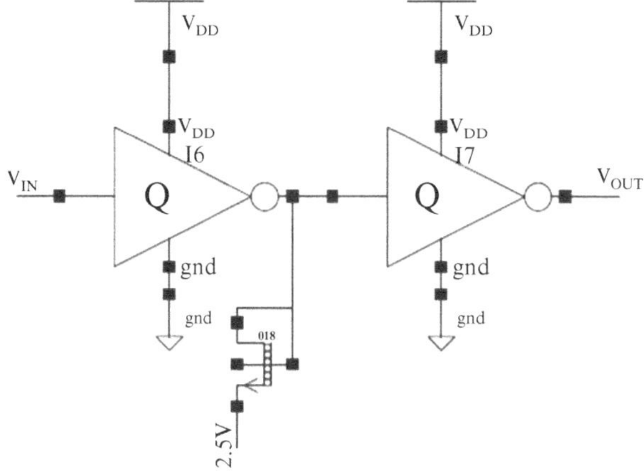

Fig. 6.17 Circuit diagram of back-to-back complement functions with stabilizing transistor

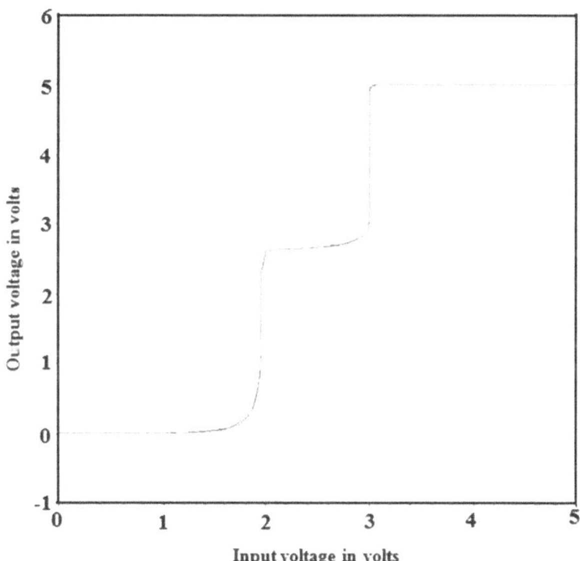

Fig. 6.18 $V_{OUT} - V_{IN}$ transfer characteristics for back-to-back QD complement functions with stabilizing transistor

It is possible to add an extra depletion-mode transistor tied to $V_{DD}/2$, so that the intermediate voltage point is fixed at $V_{DD}/2$ (Figs. 6.17 and 6.18). Note that this transistor is not necessary for proper functioning, but its use does improve noise margins slightly at the cost of an extra transistor.

A fully functional memory cell can be constructed by adding 2 transistors for data pass transistors for a total of 6 transistors (7 with stabilizing transistor). Note that this is more than the number of devices in NDR- and RTD-based three-state flip-flops [10]. However, it is significantly less than any other MOSFET-based three-state register. For example, Cilingiroglu and Ozelci use 12 transistors for their three-state memory cell [11].

References

1. http://www-device.eecs.berkeley.edu/~bsim3/bsim4_get.html
2. http://www-device.eecs.berkeley.edu/~bsim3/intro.html
3. Chandy, J.A., Jain, F.C.: Multiple valued logic using 3-state quantum dot gate FETs. In: Proceedings of International Symposium on Multiple-Valued Logic, Dallas, Texas, USA (2008)
4. Jain, F.C., Heller, E., Karmarkar, S., Chandy, J.: Device and circuit modeling using novel 3-state quantum dot gate FETs. In: Proceedings of International Semiconductor Device Research Symposium, College Park, MD, USA (2007)
5. Doyle, B., et al.: High performance fully-depleted tri-gate CMOS transistors. IEEE Electron Device Lett. **24**(4), 263–265 (2003)
6. Veendrick, H.: Short –circuit dissipation of static CMOS circuitry and its impact on the design of buffer circuits. IEEE J. Solid State Circuits **SC-19**(4), 468–473 (1984)
7. Raychowdhury, A., Roy, K.: A novel multiple-valued logic design using ballistic carbon nanotube FETs. In: Proceedings of International Symposium on Multiple-Valued Logic, Toronto, Canada, pp. 14–19 (2004)
8. Balla, P.C., Antoniou, A.: Low power dissipation MOS ternary logic family. IEEE J. Solid State Circuits **19**(5), 739–749 (1984)
9. Hanyu, T., Kameyama, M.: A 200 MHz pipelined multiplier using 1.5 V-supply multiple valued MOS current mode circuits with dual-rail source-coupled logic. IEEE J. Solid State Circuits **30**(11), 1239–1245 (1995)
10. Uemura, T., Baba, T.: A three-valued d-flip-flop and shift register using multiple junction surface tunnel transistors. In: Proceedings of International Symposium on Multiple-Valued Logic. IEEE Transactions on Electron Devices. **49**(8), 1336–1340 (2002)
11. Ilingiroglu, U.C., Ozelci, Y.: Multiple-valued static CMOS memory cell. IEEE Trans. Circuits Syst. II Analog Digit. Signal Process. **48**, 282–290 (2001)

Chapter 7
Analog-to-Digital Converter (ADC) and Digital-to-Analog Converter (DAC) Using Quantum Dot Gate Field-Effect Transistor (QDGFET)

Analog circuits such as analog-to-digital converters (ADCs) and digital-to-analog converters (DACs) using QDGFET are discussed in this chapter. This chapter also introduces another quantum dot-based device known as quantum dot gate nonvolatile memory (QDNVM) which is used as a variable threshold voltage transistor (QDVTH) to realize compact comparator circuits. This chapter discusses the reduction of the number of elements in circuits based on QDGFET. Different comparator architectures are also discussed here. The design of six-bit ADCs and DACs is also presented. Performance analysis of a three-bit ADC as well as six-bit ADC is also presented. The design of a "reconstruction circuit" which successfully reconstructs the input analog signal concludes the chapter.

7.1 Introduction

Most natural phenomena are in analog forms. Almost any measurable quantity is analog in nature, such as temperature, pressure, speed, and time. Analog-to-digital converter (ADC) and digital-to-analog converter (DAC) are of great importance for conversion between analog signal and digital signal.

In this section, we will look at methods to convert from digital signals to analog signals and vice versa using QDGFETs. An analog quantity is one that has a continuous set of values over a given range, as contrasted with discrete values for digital case.

7.2 Analog-to-Digital (A/D) Conversion

Analog-to-digital conversion is a process of converting analog quantity into its digital form. ADC is required specially for the processing of analog signal in computer to acquire more information from them. For information storage or for

S. Karmakar, *Novel Three-state Quantum Dot Gate Field Effect Transistor: Fabrication, Modeling and Applications*, DOI 10.1007/978-81-322-1635-3_7,

display, ADC converter also required to convert analog signal into their digital form. The processed signal again converts back to analog signal by digital-to-analog converter. There are several types of A/D conversion methods which are discussed in Sect. 7.2.1.

7.2.1 Existing A/D Conversion Method

7.2.1.1 Simultaneous A/D Converter

Simultaneous A/D converters are also known as flash ADC. In simultaneous A/D converter, different comparators having different reference voltages compare analog signal with their reference voltages simultaneously [1]. Since all comparator works at the same time, the other name of simultaneous ADC is flash ADC. Based on different reference voltages, different comparators produce different outputs. When the input analog signal for a comparator is more than that of its reference voltage, the comparator output is high. When the analog signal is less than the reference voltage, the comparator output is low. Figure 7.1 shows the block diagram of a flash ADC.

Primary advantage of flash ADC is that it provides fast conversion time. But the main disadvantage is the number of comparators. For a reasonable sized binary umber, large number of comparators is required.

The resistive voltage network is used to set up reference voltages for different comparators. These comparators are used for quantization of analog signal. This quantized analog signal is then fed up to the input of the following encoder which encodes different quantized levels to different digital bit combinations. The highest bit combination is assigned for highest level of quantized level, and lower levels are assigned for lower quantized voltage levels of analog signal.

The sampling rate of an A/D converter determines the accuracy of converting analog signal into its digital form [2, 3]. The accuracy increases with the number of samples. But the number of comparators as well as the number of circuit elements also increases with the number of samples. So a trade-off is necessary between the accuracy of the ADC and the number of circuit elements.

7.2.1.2 Stair-Step Ramp A/D Converter

Stair-step ramp A/D conversion [4] is another ethos to convert analog signal to its digital form. This method is also known as digital ramp or counter method. This method employs counter and digital-to-analog converter (DAC) to generate the digital output of an analog input signal. Figure 7.2 shows the diagram of a stair-step A/D converter.

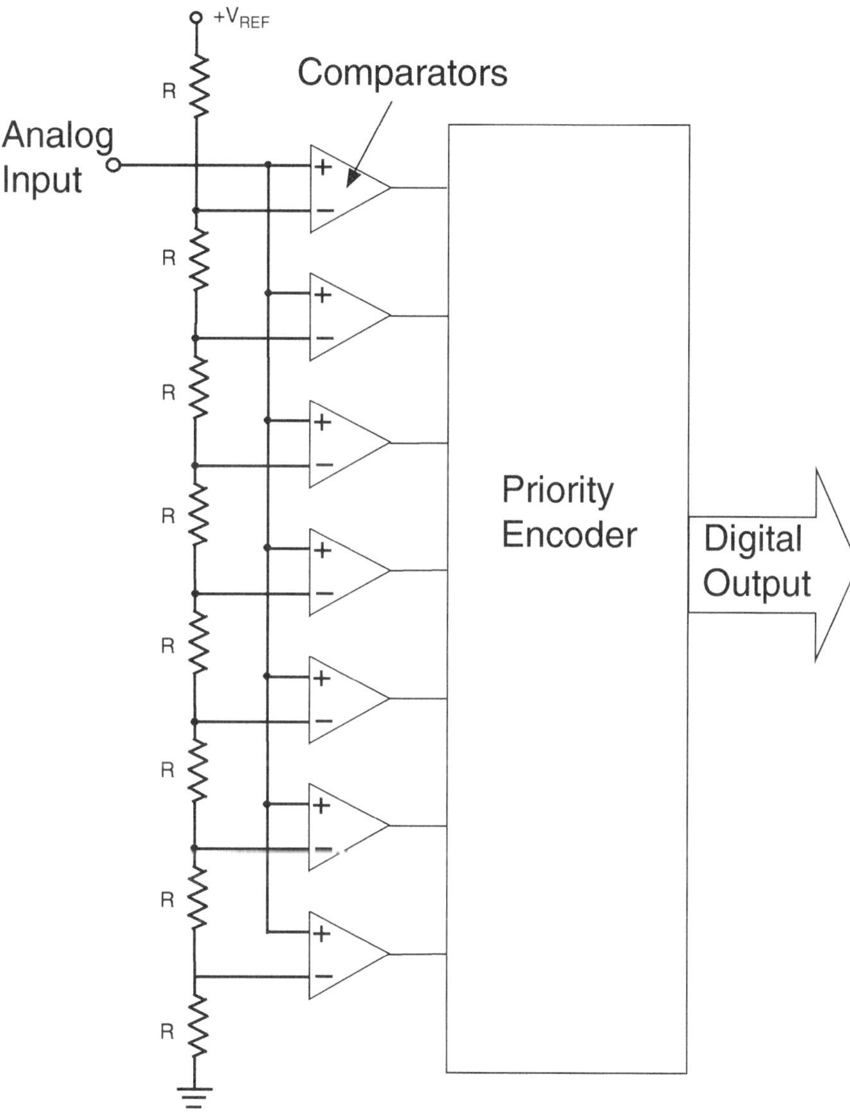

Fig. 7.1 A three-bit simultaneous analog-to-digital converter

In this A/D conversion, the counter starts from reset state and the DAC output starts from 0. When the analog input is high, the comparator output is high. The following AND gate of the comparator becomes transparent to its input clock signal and passes the input clock signal to the counter circuit. The clock pulses begin and advance the counter through its different binary states and produce a stair-step reference voltage for D/A converter.

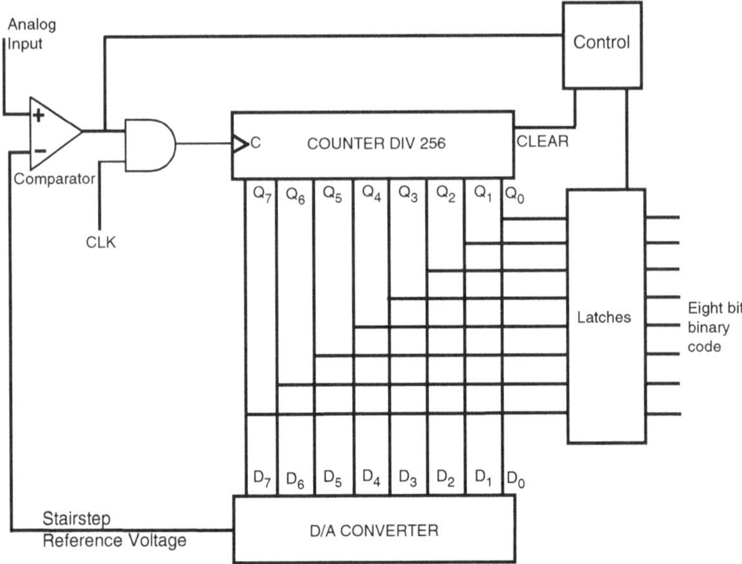

Fig. 7.2 Stair-step ramp eight-bit analog-to-digital converter

When the stair-step reference from the D/A converter output reaches to the analog input voltage level, the comparator output is low. The low output of the comparator disables the AND gate and restrains the clock signal for entering to the counter circuit, and the counter stops advancing.

The state of the counter at this point is equal to the number of steps in the reference voltage at which the comparison occurs. The binary number represents the analog input voltage. The control logic following the counter loads the binary count and resets the counter for the next sequence of count.

This method is slower than the simultaneous method or flash method. The worst can happen for maximum analog input signal. In this case, the counter needs to count the maximum number of states before conversion occurs. The conversion time varies depending on the analog input voltage.

7.2.1.3 Tracking A/D Converter

Tracking A/D converter uses an up/down counter [5, 6] instead of up converter in stair-step ADC. The counter does not need to reset after every cycle which makes it faster than the stair-step ADC. Figure 7.3 shows a typical eight-bit tracking A/D converter.

Tracking ADC converter works in a different way than stair-step ADC. When the D/A output is less than the analog input signal, the comparator output is high, the counter is in up mode and produces up sequence until the D/A converter output

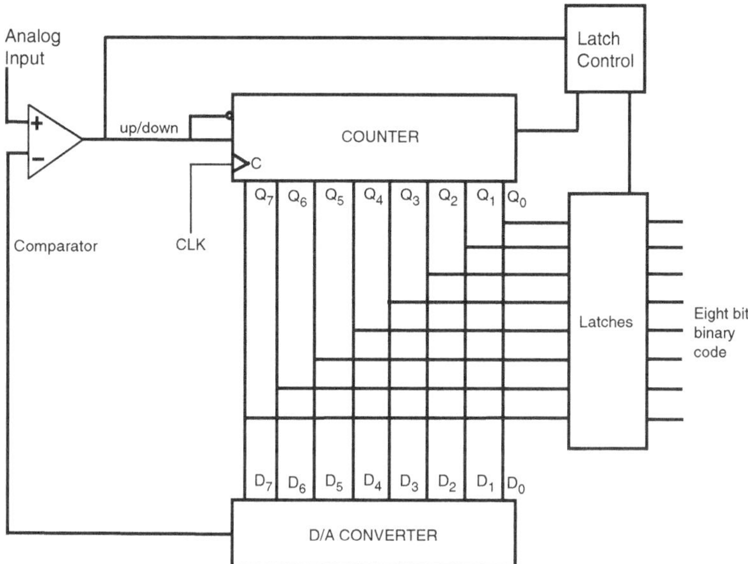

Fig. 7.3 Eight-bit tracking analog-to-digital converter

voltage equals to the analog input voltage. When the D/A output equals to the analog input voltage, the comparator output is low and counter works in down mode, and down sequence starts.

If the input analog signal goes downward, the counter follows the input analog signal until the D/A output lowered to the analog signal value and the comparator output becomes zero at this point. The control circuit stores the equivalent value to the lath. When the input is a constant, the comparator output becomes high and low in successive cycle, and the counter output causes an oscillation between two binary states.

7.2.1.4 Single-Slope A/D Converter

The basic disadvantage of a stair-step ramp A/D converter and tracking A/D converter is the implementation of D/A converter in the circuit architecture. The disadvantage can be removed in single-slope A/D converter. This type of converter does not require a D/A converter. It uses a linear ramp generator instead of D/A converter to produce a constant slope ramp voltage. The circuit diagram for a single-slope A/D converter is shown in Fig. 7.4.

At the beginning of the conversion cycle, the counter is in reset state and the ramp generator output also 0 V. At this stage, the analog input is greater than the reference voltage and the output of the comparator is high and which enables the counter and starts up the ramp generator.

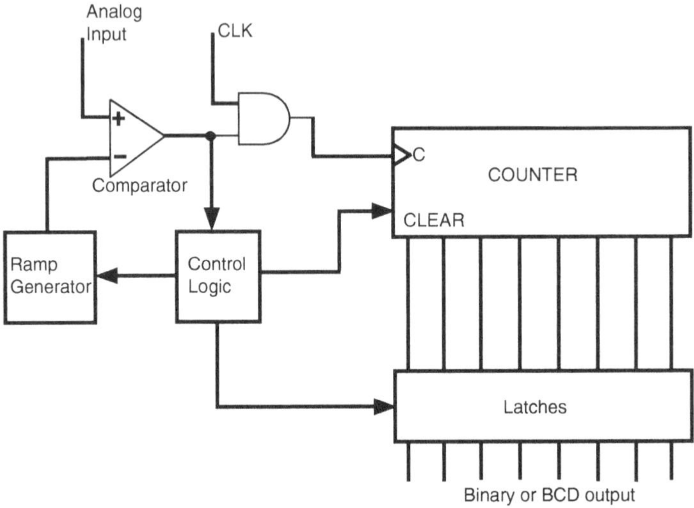

Fig. 7.4 Eight-bit single-slope analog-to-digital converter

The ramp generator [7, 8] will increase until its output equals to the analog input voltage. When ramp generator output equals to the analog input voltage, the comparator output is low, and the counter as well as the ramp generator stops working.

The equivalent value of the analog signal is stored in the latch by the control circuit.

7.2.1.5 Dual-Slope A/D Converter

Dual-slope A/D converter works in the same basic principle of single-slope A/D converter except that a variable slope ramp and a fixed slope ramp are used.

7.2.1.6 Successive-Approximation A/D Converter

One of the most popular A/D converter is successive-approximation A/D converter [9]. Conversion time for this type of A/D converter is less than any other method except simultaneous method. The conversion time is fixed for any analog input signal.

The basic block diagram of a successive-approximation A/D converter is shown in Fig. 7.5. The block diagram consists of a D/A converter, a successive-approximation register (SAR), and a comparator. Initially, D/A converter output is zero, and comparator output is high because of finite analog voltage.

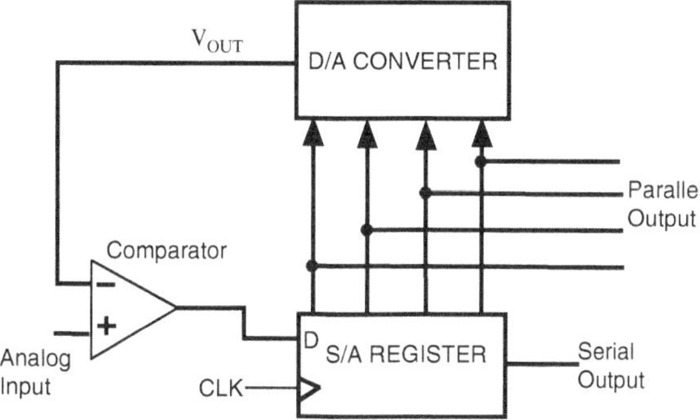

Fig. 7.5 Successive-approximation analog-to-digital converter

The input bits in D/A converter enable one at a time, starting with the most significant bit (MSB) in each clock cycle. The output of D/A converter is compared with the analog input signal. Until the D/A converter output is lower than the analog input, the D/A converter moves on. When D/A converter output is more than the analog input, comparator output is low which causes the register bits to reset. Until D/A converter output is lower than analog input voltage, the bit is retained in the register.

The conversion in this A/D converter starts from with the MSB first, then the next most significant bit, then the next, and so on. After all the bits of the D/A have been tried, the conversion cycle is complete.

7.2.2 Variable Threshold Voltage Transistor

The variable threshold voltage transistor has a similar structure as a three-state quantum dot gate FET except the control gate insulator is on the top of the quantum dot layer in the gate region as shown in Figs. 7.6 and 7.7. Based on the applied gate voltage, electrons tunnel from the inversion channel between the source and drain to the quantum dot layer on the gate region and are stored there because of the control gate insulator layer on top of gate region. The amount of stored charge in the quantum dots layer varies the threshold voltage of the device. The variable threshold voltage transistor can also be used as a nonvolatile memory [10, 11]. In the variable threshold voltage transistor (QDVTH), the threshold voltage of the device varies based on the charge stored in the quantum dot layer in the gate region.

7.2.3 Comparator

The basic building block of an analog-to-digital converter (ADC) and a digital-to-analog converter (DAC) is the comparator. Figure 7.8 shows a comparator circuit

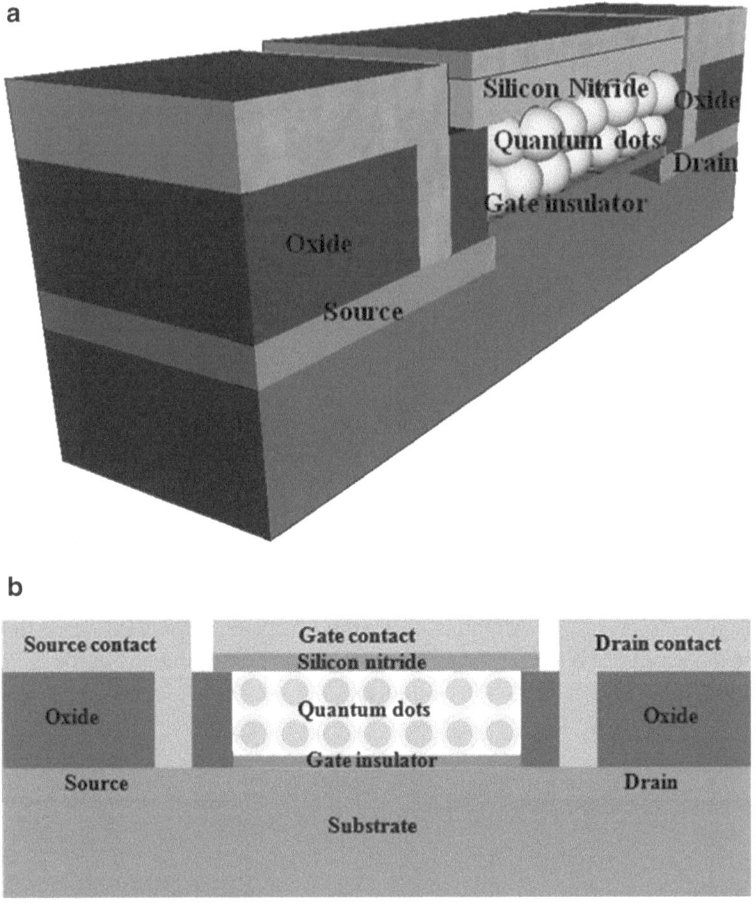

Fig. 7.6 (**a**) Device structure and (**b**) cross-sectional schematic of a QDVTH

based on a variable threshold voltage transistor. In this circuit the variable threshold voltage transistor is used as a voltage dependent resistor. The drain-to-source voltage of the variable threshold voltage transistor changes the source voltage of the pull-down NMOS of the comparator, which effectively changes the gate-to-source voltage of the pull-down NMOS transistor and the crossover point of the comparator circuit. In this circuit, the variable threshold voltage transistor is used as a reference voltage source in the source terminal of the pull-down NMOS FET. In essence, the variable threshold voltage transistor changes the switching voltage or the reference voltage of the comparator based on its threshold voltage (Fig. 7.9).

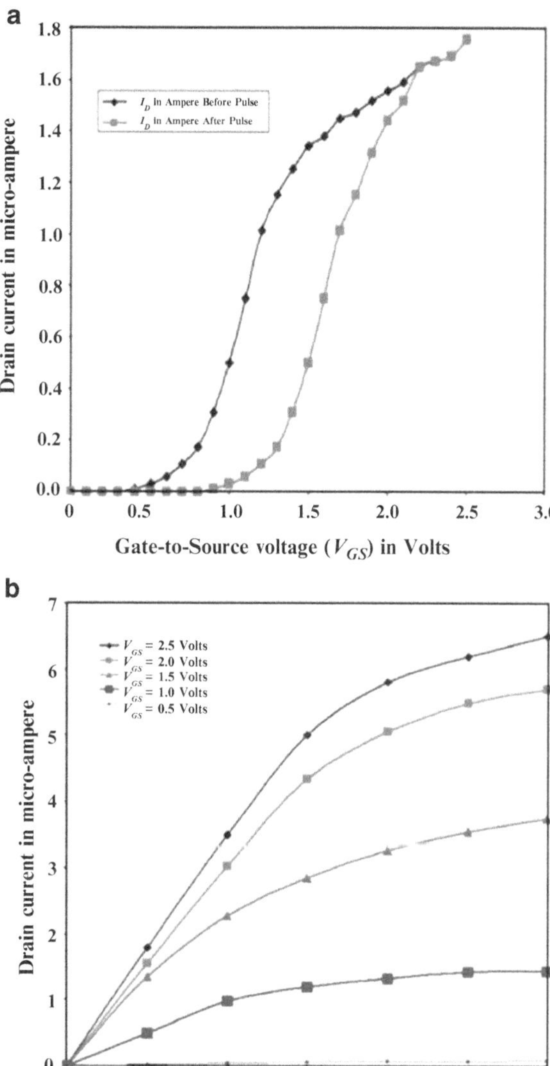

Fig. 7.7 (**a**) Threshold voltage shift of a QDNVM having SiOx-cladded Si quantum dots on top of gate insulator [17] and (**b**) output characteristics of the QDNVM

7.2.4 QDGFET-Based Three-Bit Analog-to-Digital Converter (ADC)

A flash ADC is the simplest and fastest ADC architecture. The basic building block of a flash ADC architecture is shown in Fig. 7.10. The architecture is composed of two building blocks: comparator and encoder. The comparator is used to quantize

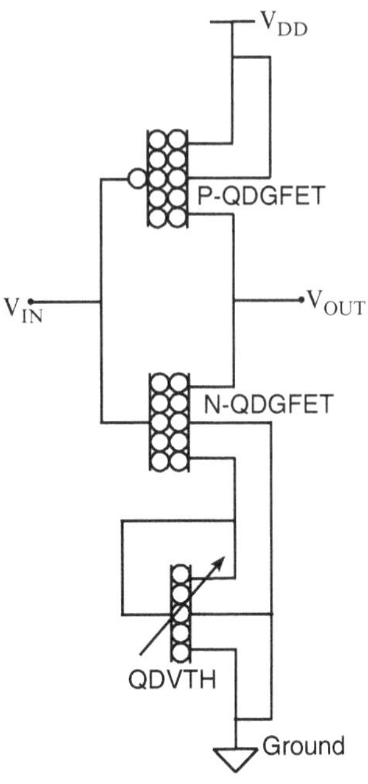

Fig. 7.8 Comparator circuit based on QDGFETs

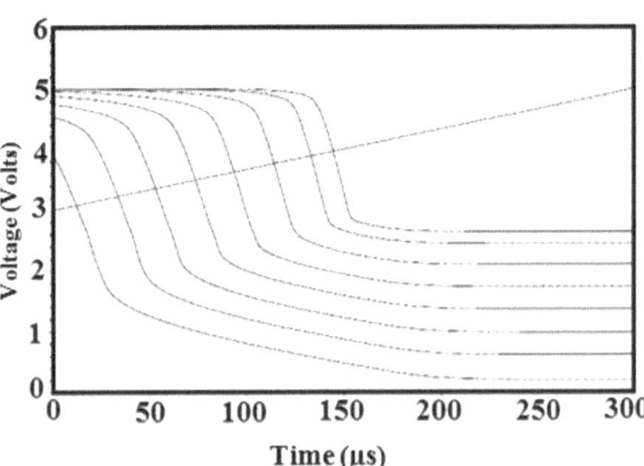

Fig. 7.9 Changes of switching voltage of comparator based on reference voltage

the analog input signal based on the required number of levels, and the encoder is used to convert these into different digital bit combinations.

The comparator circuit is designed using quantum dot gate FETs and a variable threshold voltage transistor. As discussed in Sect. 7.2.3, we set the reference voltages of different comparators based on our required voltage level for quantization of the input analog voltage. Eight comparators divide the analog input voltage in eight different quantized levels. The encoder comprises of two

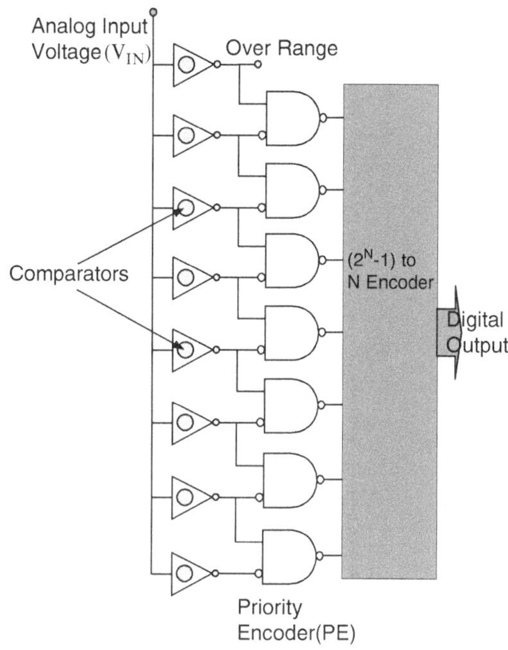

Fig. 7.10 Three-bit flash ADC architecture based on QDGFET

basic building blocks. The first block encodes the difference of two consecutive input quantized signals (Fig. 7.11). The 8-to-3 encoder converts these differences into different bit combinations. Figure 7.12 shows the output waveform of the three-bit ADC.

7.3 Three-Bit Digital-to-Analog Converter (DAC)

Digital-to-analog conversion is an important interface process in many applications. An analog signal that has been digitized for processing or transmission must be changed back into an approximation of the original signal to drive other analog devices.

7.3.1 Existing D/A Converter

7.3.1.1 Binary-Weighted Input D/A Converter

Binary-weighted D/A **converter** uses a resistor network where the resistor values represent the binary weights [12] of each input bits of the input digital code.

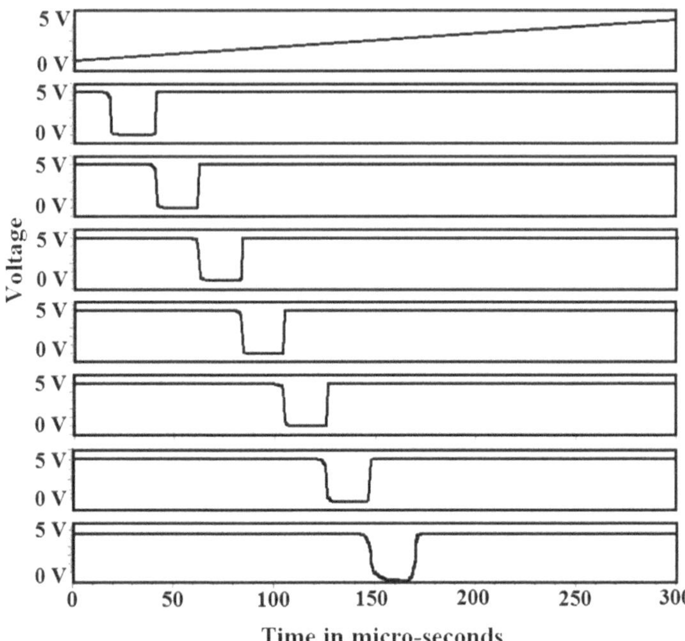

Fig. 7.11 Encoded output of the difference of two consecutive signals from priority encoder

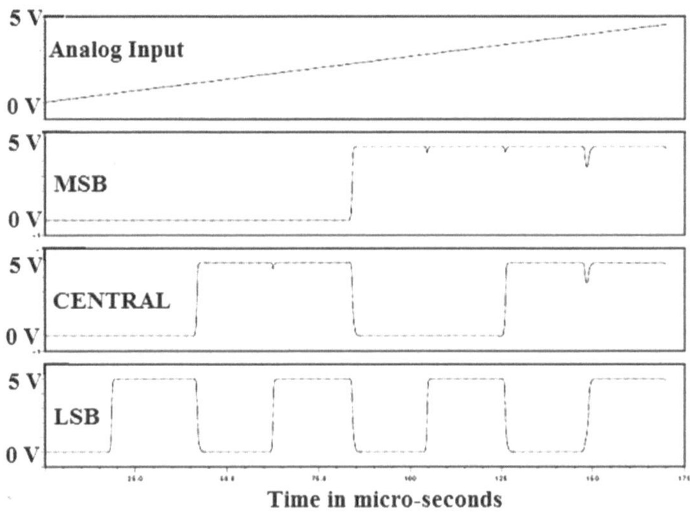

Fig. 7.12 Input-output waveforms of the designed A/D converter

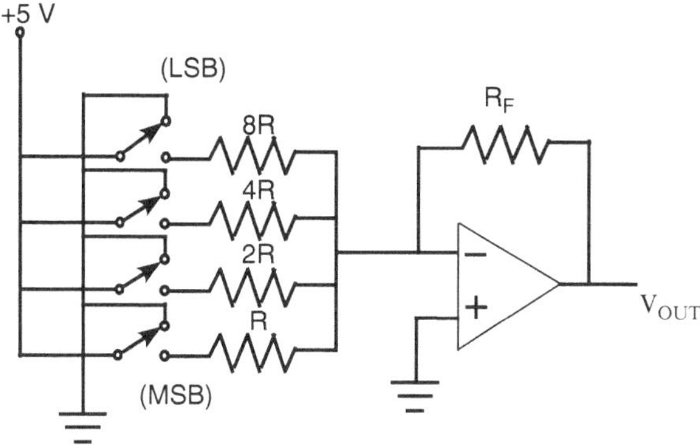

Fig. 7.13 Four-bit binary-weighted D/A converter

Figure 7.13 shows a four-bit binary-weighted D/A converter. In this circuit switches are required to input different bit combinations in the circuits. The switches are followed by high-impedance operational amplifier (OPAMP). The OPAMP is connected as inverting an amplifier. The non-inverting input of this OPAMP is ground. The inverting input of this OPAMP behaves like a virtual ground so that all current flows through the feedback resistor R_F and the output of the OPAMP is proportional to the feedback resistor.

The circuit operation can be explained as follows. The values of different resistors are chosen based on the weight of the binary input. The lowest value of the resistor is assigned for highest binary input so that the highest current flows through this resistor for the highest value of binary input. Other resistors are multiples of R: 2R, 4R, and 8R, corresponding to the binary weights 2^2, 2^1, and 2^0, respectively. The highest value resistor for lowest binary digit allows less current in the feedback resistor for lowest bit in the digital code. One of the main disadvantages of binary-weighted D/A converter is the number of different resistors ranging having values from R to $(2^N - 1)$ R having tolerances in one part of $(2^N - 1)$ (less than 0.5 %) to accurately convert the input, which makes this type of D/A converter very difficult for mass production [13, 14].

7.3.1.2 R/2R Ladder D/A Converter

R-2R ladder is another method of D/A conversion [15]. The resistor problem in binary-weighted D/A converter is absent in this D/A converter. Figure 7.14 shows the block diagram of an R-2R ladder D/A converter.

In this D/A converter, only two combinations of resistors are required for digital-to-analog signal conversion. In this circuit, each successive lower bit combination produces output analog voltage that is halved the previous one and makes the output voltage proportional to the binary weight of the input bits [16].

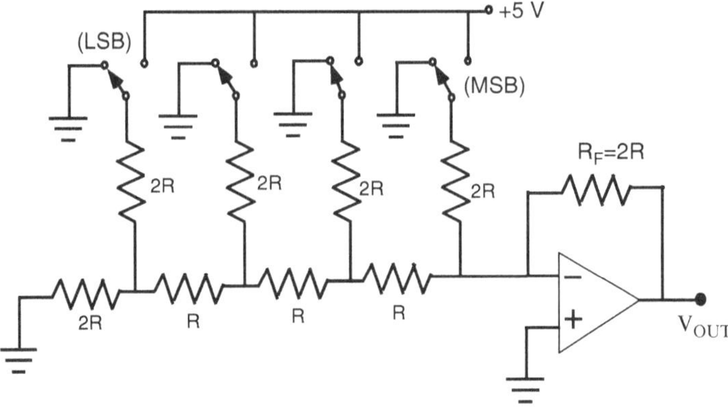

Fig. 7.14 Four-bit R-2R ladder D/A converter

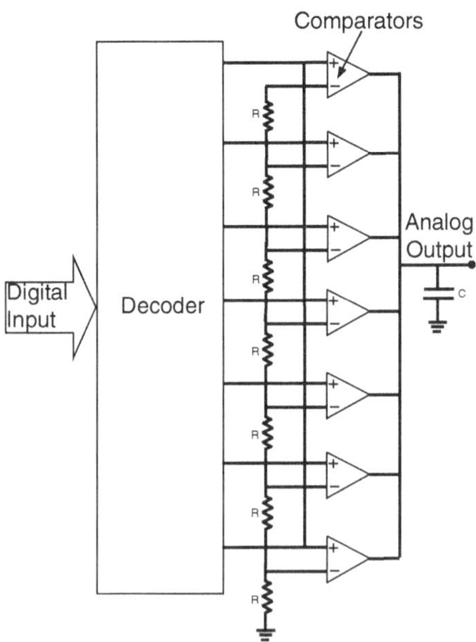

Fig. 7.15 Flash architecture of D/A converter

7.3.2 D/A Converter: Flash Architecture

A flash digital-to-analog converter (DAC) architecture is shown in Fig. 7.15.

Figure 7.16 shows our designed architecture for a digital-to-analog converter (DAC) based on quantum dot gate FETs (QDGFET). In this architecture, the three-to-eight decoder decodes the input digital bit combinations and activates the corresponding output.

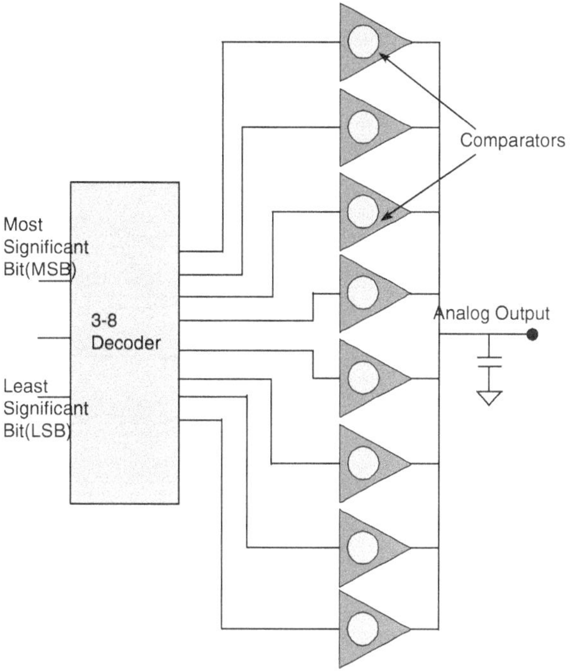

Fig. 7.16 D/A converter using QDGFET

Each output lines from the decoder activates the corresponding comparators. The output capacitor discharges through one comparator at a time based on the reference voltage of the comparator which depends on the threshold voltage of variable threshold voltage transistor of the comparator. The designed circuit produces higher reference voltage for higher bit combinations. The output of the designed three-to-eight decoder and the digital-to-analog converter (DAC) are shown in Figs. 7.17 and 7.18, respectively.

7.4 Noise Analysis

For an ADC, the ideal SNR should be 6.02 N + 1.76 dB where N is the number of bits [17]. The 8,192 point FFT of our designed three-bit ADC circuits is shown in Fig. 7.19. The noise is calculated using the measured magnitude of the bins in Fig. 7.19. The bins corresponding to the fundamental input signal and its multiple frequencies (or harmonics) are excluded from the denominator of SNR calculation. The converter's rms noise is measured by taking the magnitude of each of these bins, squaring it, adding all of the squared bins together, and then calculates the square root of that summation. This analysis shows an SNR of 18 dB for our designed ADC circuits, which is close to the ideal value of 19.82 dB.

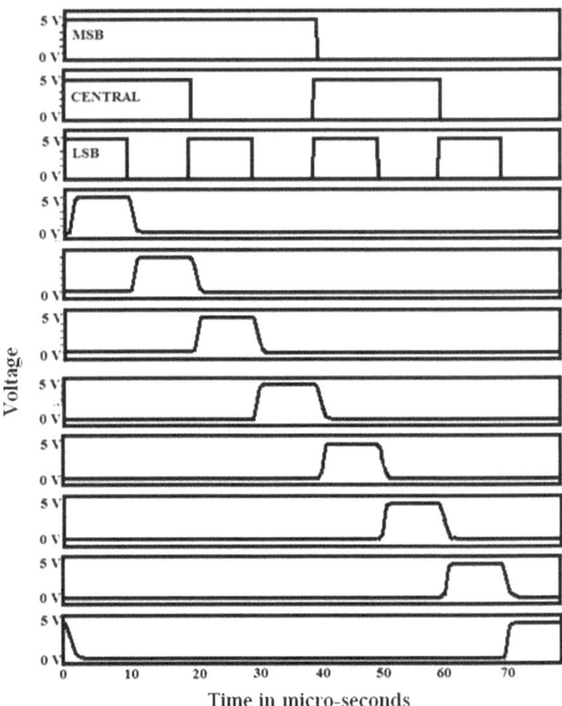

Fig. 7.17 Input-output waveforms of three-to-eight decoder output

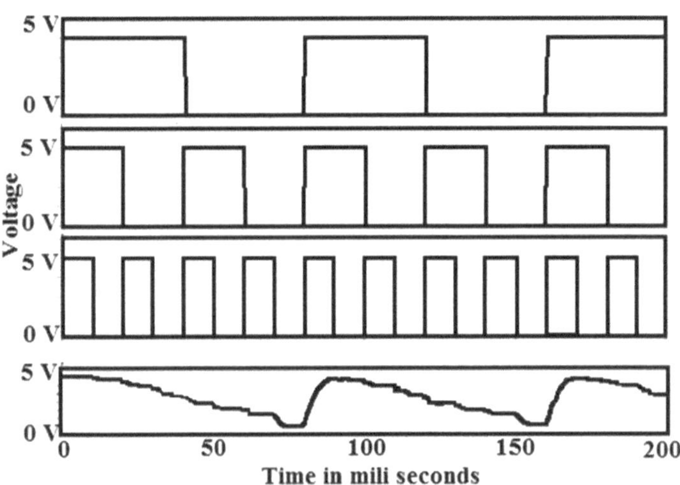

Fig. 7.18 Input-output waveforms of the D/A converter

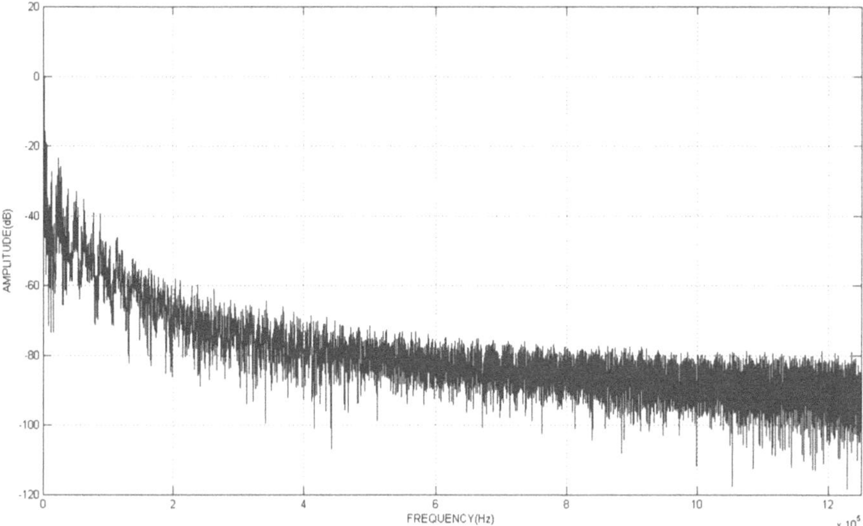

Fig. 7.19 8,192-point FFT of the D/A converter output

Table 7.1 Different performance parameters of the three-bit ADC

Signal-to-noise ratio (SNR)	18.2348 dB
Signal-to-noise and distortion ratio (SNDR)	12.7508 dB
Total harmonic distortion (THD)	−14.1947 dB
Spurious frequency dynamic range (SFDR)	16.0453 dBFS
Error in number of bit (ENOB)	1.82571 bits

In addition to the SNR calculation, we also calculate the signal-to-noise and distortion ratio (SNDR), total harmonic distortion (THD), spurious frequency dynamic range (SFDR), and error in number of bit (ENOB) using 8,192-point FFT of our designed three-bit ADC circuits shown in Fig. 7.19 (Table 7.1).

7.5 Six-Bit Analog-to-Digital Converter (ADC)

7.5.1 Comparator Design

The basic building block of an analog-to-digital converter (ADC) and a digital-to-analog converter (DAC) is the comparator. The comparator design using static CMOS architecture is already mentioned in the previous section of three-bit analog-to-digital (ADC) circuit design. In designing the six-bit ADC, the number of transistor also increases. Ratioed logic is an attempt to reduce the number of transistors to implement a given logic function, often at the cost of reduced robustness

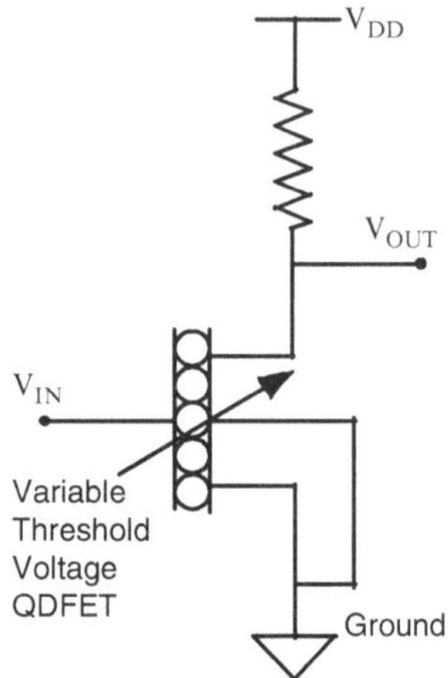

Fig. 7.20 Circuit diagram of a comparator based on QDNVM

and extra power dissipation. The static power dissipation of pseudo-NMOS limits its use. When area is most important, however, its reduced transistor count compared to complementary CMOS is quite attractive. Figure 7.20 shows a comparator circuit based on QDNVM, which acts as a variable threshold voltage transistor (QDVTH). In this circuit, the QDVTH is used as a reference voltage source in the pull-down function of comparator. The QDNVM or the QDVTH turns on based on its threshold voltage which depends on the stored charge in the gate region. When the input is less than the threshold voltage of the QDNVM, it is off and the output of the comparator is high. When the input is more than the threshold voltage of the QDNVM, it is on and output is low. In essence, the variable threshold voltage transistor changes the switching voltage or the reference voltage of the comparator based on its threshold voltage (Fig. 7.21). The new design of the comparator circuit decreases the number of transistors per comparator from 3 in the previous architecture to 1 in new architecture.

7.5.2 ADC Architecture

Similar to three-bit analog-to-digital converter (ADC), six-bit flash ADC also comprises of comparators and encoder circuits. In this section, we design the comparator circuit based on QDNVM which acts as a variable threshold voltage transistor. As discussed in the previous paragraph, we set the reference voltages of different comparators based on our required voltage level for quantization of the input analog voltage. Sixty-four comparators divide the analog input voltage in 64 different quantized levels. The encoder comprises of two basic building blocks: priority encoder and 64-to-6 encoder. The priority encoder encodes the difference of two consecutive quantized signals. The 63-to-6 encoder converts

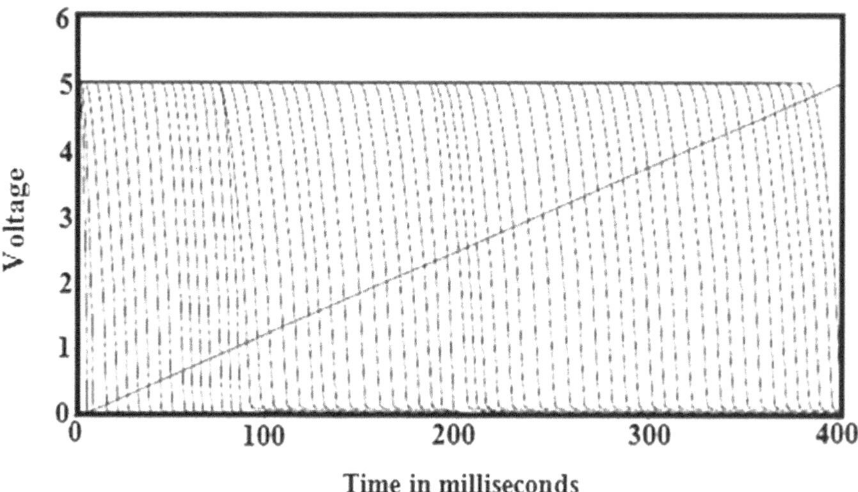

Fig. 7.21 Crossover point change with the change of threshold voltage of QDNVM

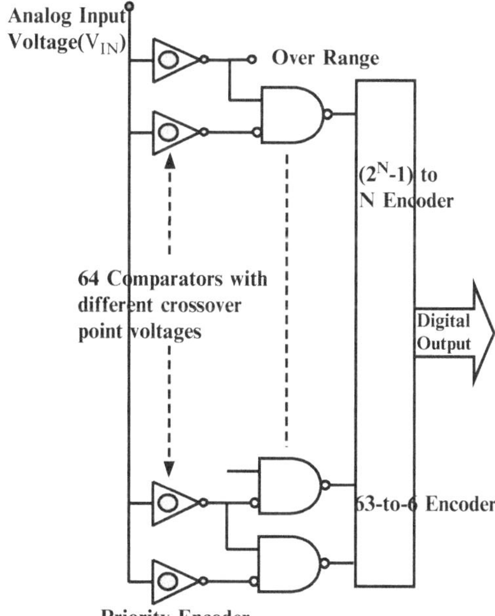

Fig. 7.22 Flash architecture of six-bit analog-to-digital converter (ADC)

these differences into different bit combinations. Figure 7.22 shows the block diagram of a six-bit analog-to-digital converter (ADC), and Fig. 7.23 shows the input and output waveform of the designed six-bit analog-to-digital converter (ADC) circuit.

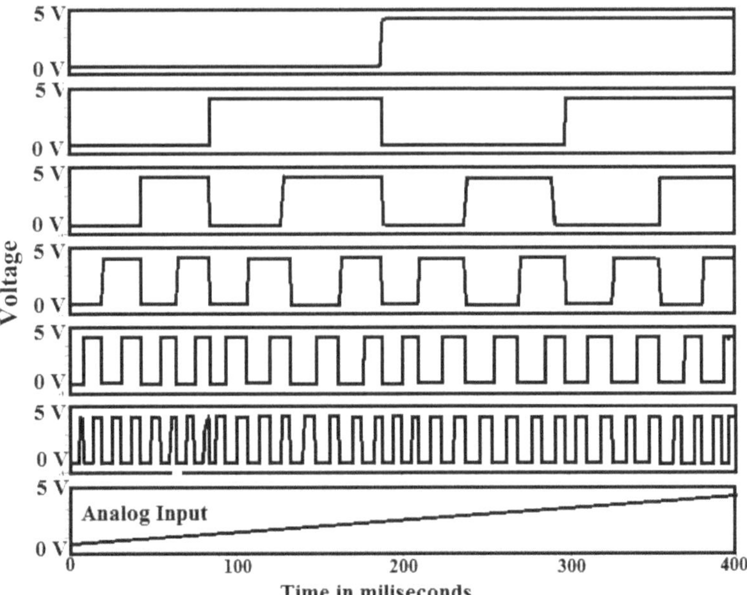

Fig. 7.23 Input-output waveforms of designed six-bit ADC

7.6 Six-Bit Digital-to-Analog Converter (DAC)

The digital-to-analog converter (DAC) has been designed following the reverse architecture of ADC circuits. Here, 6-to-64 decoder decodes six-bit digital input code and activates one output (high) at a time based on different input bit combinations.

The decoder outputs are connected to different comparator circuits having various crossover points based on corresponding QDNVM threshold voltages. Based on the decoder output, one comparator is activated at a time and produces output voltage based on crossover point voltage of the corresponding comparator. The comparator corresponding to highest bit combination produces highest voltage where as that one for lowest bit combination produces lowest voltage. Other comparators produce different voltages based on their positions in the decoder output. Figure 7.24 shows the six-bit DAC architecture, and Fig. 7.25 shows the input and output waveform of the designed six-bit DAC circuit.

7.7 Reconstruction Circuit

The block diagram for reconstruction circuit comprising six-bit ADC and DAC is shown in Fig. 7.26. Here, input sinusoidal wave is digitized by the six-bit ADC, and the digitized output of the ADC is fed to the DAC and reconstructed to the input

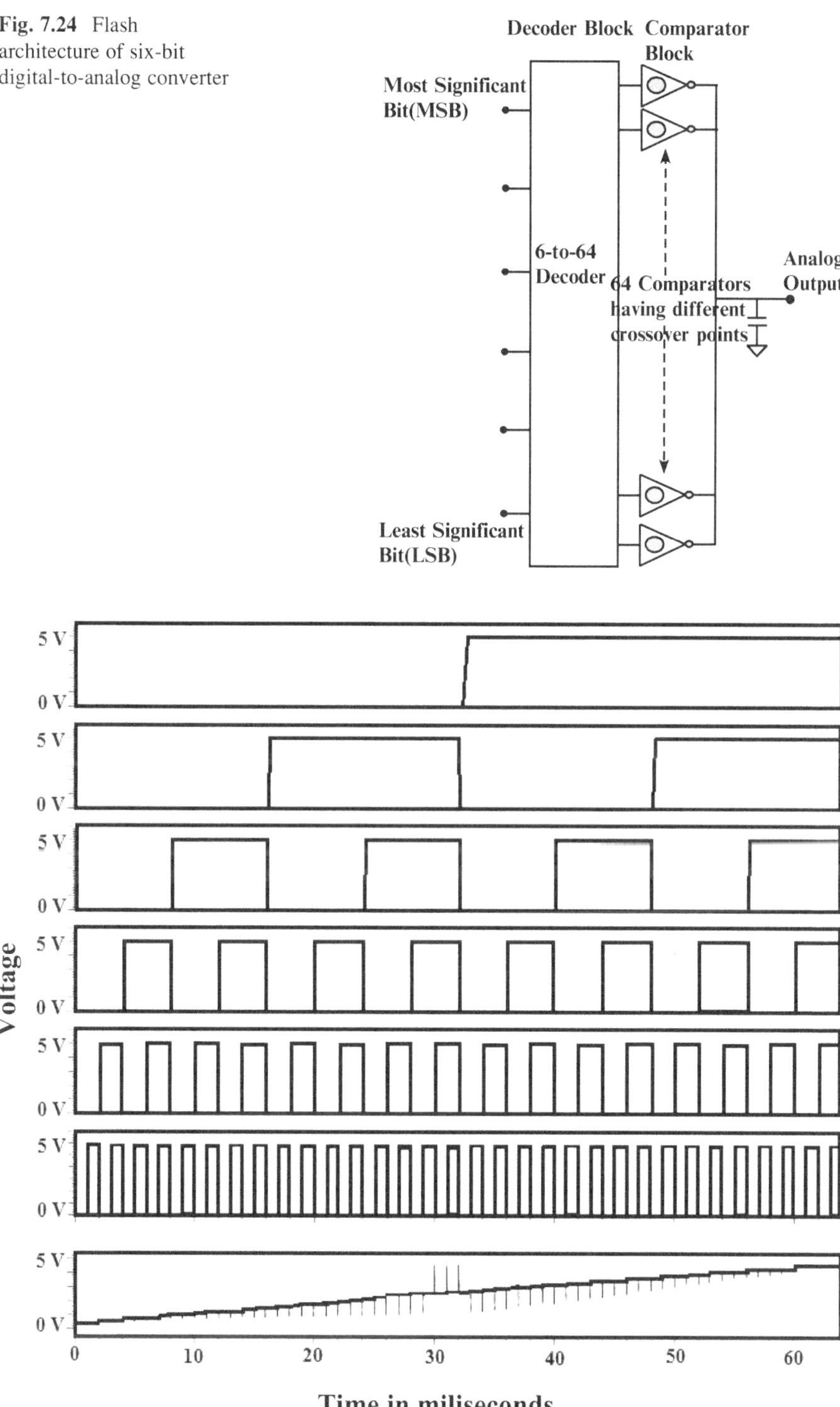

Fig. 7.24 Flash architecture of six-bit digital-to-analog converter

Fig. 7.25 Input-output waveforms of designed six-bit DAC

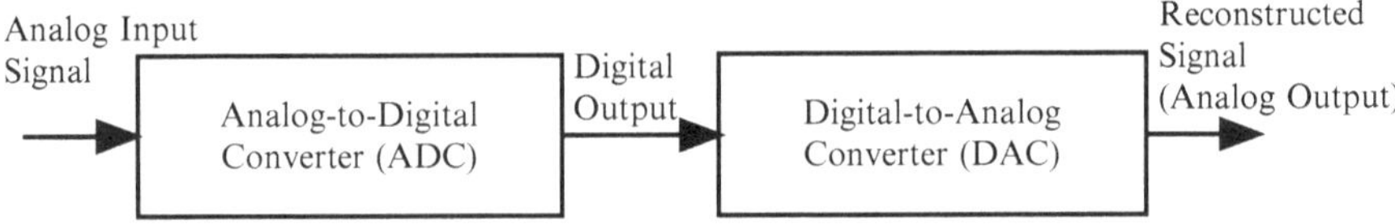

Fig. 7.26 Block diagram of reconstruction circuit

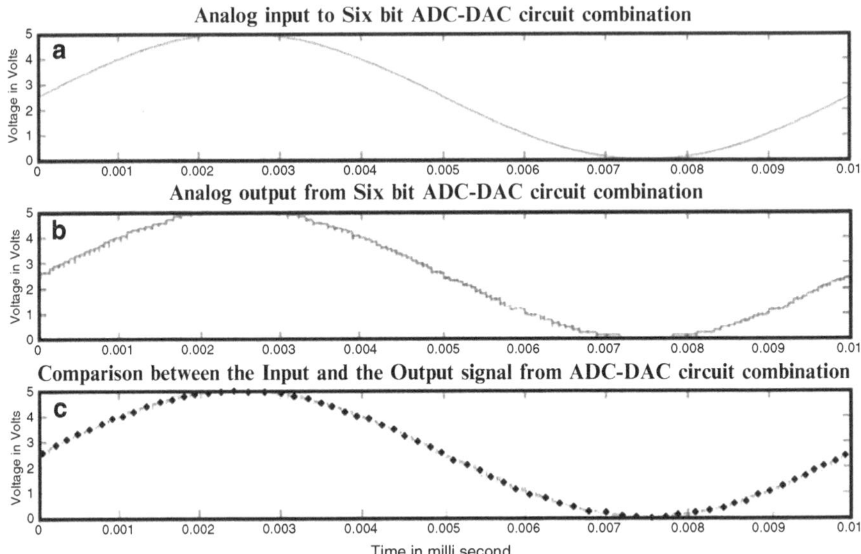

Fig. 7.27 Input (**A**) vs Output (**B**) waveform of the reconstruction circuit. Waveform (**C**) shows the comparison between input and output waveforms [17]

analog signal. The input-output waveform of the reconstruction circuit is shown in Fig. 7.27.

7.8 Noise Analysis

Signal-to-noise ratio (SNR), signal-to-noise and distortion ratio (SNDR), total harmonic distortion (THD), spurious frequency dynamic range (SFDR), and error in number of bit (ENOB) are also calculated using an 8,192-point FFT shown in Fig. 7.28. Table 7.2 shows the values of different performance parameters of designed analog-to-digital converter (ADC). For an ideal ADC, the SNR should be 6.02 N + 1.76 dB where N is the number of bits. Our designed ADC produces SNR around 37.36 dB at 100 Hz which is almost equal to ideal one. As the frequency increases, SNR decreases. At 1 MHz, the SNR is 31.52 which is less

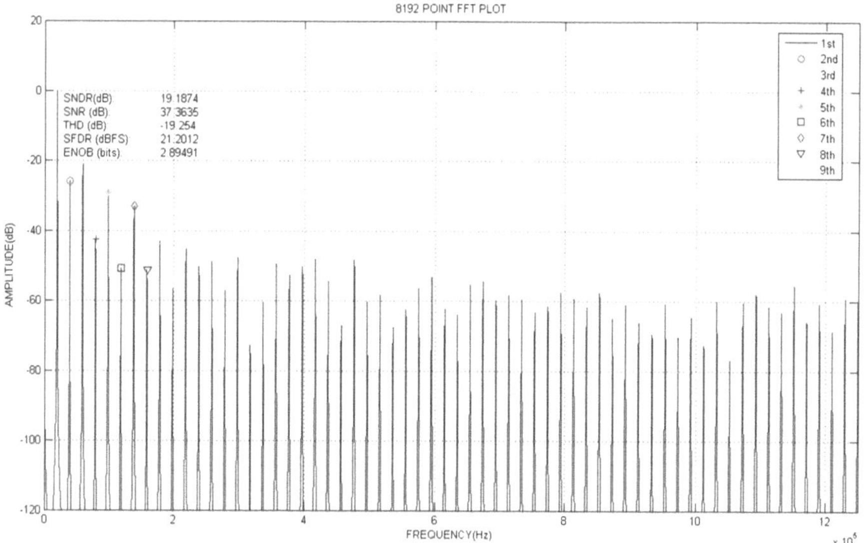

Fig. 7.28 8,192-point FFT of six-bit analog-to-digital converter (ADC) [17]

Table 7.2 Different performance parameters of the designed six-bit ADC

Signal-to-noise ratio (SNR)	37.3635 dB
Signal-to-noise and distortion ratio (SNDR)	19.1874 dB
Total harmonic distortion (THD)	−19.254 dB
Spurious frequency dynamic range (SFDR)	21.2012 dBFS
Error in number of bit (ENOB)	2.89491 bits

than a five-bit ADC. The highest operating frequency of our designed ADC is 1 MHz. Precise control of threshold voltages of different QDNVMs can produce better performance ADC.

References

1. Ferragina, V., Ghittori, N., Maloberti, F.: Low-power 6-bit flash ADC for high-speed data converters architectures. In: The Proceedings IEEE International Symposium on Circuits and Systems – ISCAS, Island of Kos, Greece (2006)
2. El-Chammas, M., Murmann, B.: A 12-GS/s 81-mW 5-bit time-interleaved flash ADC with background timing skew calibration. IEEE J. Solid-State Circuits **46**(4), 838–847 (2011)
3. El-Chammas, M., Murmann, B.: A 12-GS/s 81-mW 5-bit time-interleaved flash ADC with background timing skew calibration. In: Symposium on VLSI Circuits Digital, Honolulu, June 2010, pp. 157–158
4. Dickinson, A.R.: Device to manifest an unknown voltage as a numerical quantity. U.S. Patent 2,872,670, filed 26 May 1951, issued 3 Feb 1959

5. Barney, K.H.: Binary quantizer. U.S. Patent 2,715,678, filed 26 May 1 950, issued 16 Aug 1955
6. Gordon, B.M., Talambiras, R.P.: Information translating apparatus and method. U.S. Patent 2,989,741, filed 22 July 1955, issued 20 June 1961
7. Jimin Cheon Han, G.: Noise analysis and simulation method for a single-slope ADC with CDS in a CMOS image sensor. IEEE Trans. Circuits Syst. **55**(10), 2980–2987 (2008)
8. Lim, S., Lee, J., Kim, D., Han, G.: High-speed CMOS image sensor with column-parallel two-step single-slope ADCs. IEEE Trans. Electron Devices **56**(3), 393–398 (2009)
9. Zhimin, Z., Bedabrata, P., Fossum, E.R.: CMOS active pixel sensor with on-chip successive approximation analog-to-digital converter. IEEE Trans. Electron Devices **44**(10), 1759 (1997)
10. Karmakar, S., Chandy, J.A., Jain, F.C.: Design of ADCs and DACs using 25 nm quantum dot gate FETs. In: Lester Eastman Conference on High Performance Devices, Troy, NY, USA, 3–5 Aug 2010
11. Jain, F., Karmakar, S., Alamoody, F., Suarez, E., Gogna, M., Chan, P.-Y., Chandy, J., Miller, B., Heller, E.: Quantum dot gate 3-state InGaAs FETs for multi-valued logic circuits and advanced ADCs. In: Proceedings of Government Microcircuit Applications & Critical Technology Conference, 2010 (GOMACTech-10), Reno, NV, USA, 22–25 Mar 2010
12. Ozalevli, E., Lo, H.-J., Hasler, P.E.: Binary-weighted digital-to-analog converter design using floating-gate voltage references. IEEE Trans. Circuits Syst. **55**(4), 990–998 (2008)
13. Pastoriza, J.J.: Solid state digital-to-analog converter. U.S. Patent 3,747,088, filed 30 Dec 1970, issued 17 July 1973
14. Hnatek, E.R.: A User's Handbook of D/A and A/D Converters, pp. 282–295. Wiley, New York (1976). ISBN 0-471-40109-9
15. Lee, T.-C., Lin, C.-H.: Nonlinear R-2R transistor-only. IEEE Trans. Circuits Syst. **57**(10), 2644–2653 (2010)
16. Lee, T.-C., Lin, C.-H.: Modeling R-2R segmented-ladder DACs. IEEE Trans. Circuits Syst. **57**(1), 31–43 (2010)
17. Karmakar, S., Chandy, J.A., Jain, F.C.: Implementation of ADC and DAC using quantum dot gate non-volatile memory. J. Signal Process. Syst. June 2013. doi: 10.1007/s11265-013-0789-4

Chapter 8
Performance in Sub-25-nm Range: Circuit Model, Ternary Logic Gates and ADC/DAC

This chapter focuses on the performance of QDGFETs in the sub-nm range. This chapter focuses on the implementation of different ternary logic gates including inverter, NAND, NOR, and XOR. This chapter also discusses the universal property of ternary logic NAND and NOR gates. Noise margins, power dissipation, and different delays are also discussed here. Three-bit analog-to-digital converters (ADCs) and digital-to-analog converters (DACs) in sub-25-nm range are also simulated. Decoders based on PTI and NTI are also simulated in this chapter. Simulation results for two different kinds of decoders are presented in this chapter. The conversion from ternary logic to binary logic is also presented.

8.1 QDGFET Circuit Model for Sub-25-nm Range

A detailed circuit model for quantum dot gate FET (QDGFET) has already been discussed in Chap. 6. In this chapter, the circuit model is extended for sub-25-nm feature size. To extend the long channel model to sub-25-nm model, Predictive Technology Model (PTM) is used to extract the parameters and use them in BSIM model. Figure 8.1 shows the transfer characteristic (I_D vs. V_{GS}) of the modeled QDGFET. The presence of intermediate state even in the sub-25-nm circuit model will help to design ternary logic circuits based on this device in the sub-25-nm range. This chapter discusses the design of different logic circuits based on QDGFET in sub-25-nm range.

8.2 Scaling the Supply Voltage

Technology scaling also forces the supply voltage to scale down at a similar rate as device dimensions. But the device threshold voltage remains almost constant. It does not depend on scaling that much. With the reduction of supply voltage, the

S. Karmakar, *Novel Three-state Quantum Dot Gate Field Effect Transistor: Fabrication, Modeling and Applications*, DOI 10.1007/978-81-322-1635-3_8,

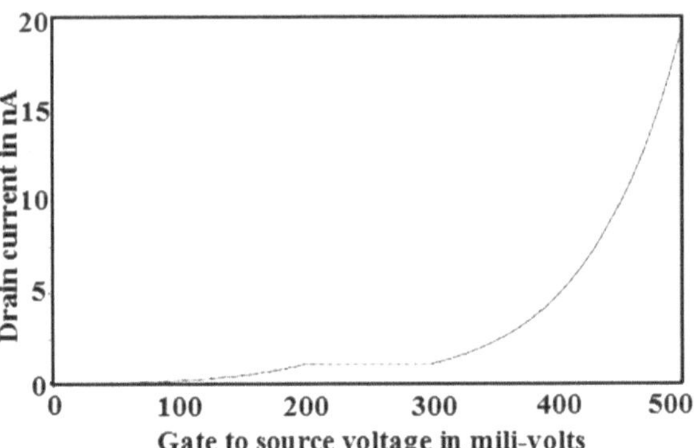

Fig. 8.1 Transfer characteristics of a QDGFET having 25-nm feature size

gain of the inverter also increases. Instead of this improvement in the low supply voltage in the dc characteristics of the device, there are so many disadvantages associated with the reduction of the supply voltage for digital circuit operation.

- The energy dissipation in any circuits decreases with the reduction of supply voltages [1, 2]. But this positive impact on energy dissipation is dominated by the increase in delay of the gate at reduced supply voltages.
- When the supply voltage becomes comparable with different intrinsic voltages levels of the device, the device operation becomes sensitive with different device parameters [3] such as threshold voltage, drain-induced barrier lowering (DIBL), and gate-induced drain leakage (GIDL).

Signal swing decreases with the supply voltage [4, 5], which decreases the internal noise between different devices in the internal circuit architecture. But on the other hand, the device becomes more sensitive to the external noise sources because they do not scale.

Because of subthreshold operation of the transistor, even at low supply voltage, inverter characteristics are attainable even though the supply voltage is not large enough to turn the transistor on. Subthreshold current can switch the gate between low and high labels and produce enough gain to produce acceptable voltage-transfer characteristic (VTC). Lower value of switching current enforces the gate a very slow operation. Besides this, as the subthreshold current increases for low feature size, the basic switching property of the transistor is gradually diminished.

With the decrease of supply voltage, V_{OL} and V_{OH} are no longer at the supply rails, which decreases the noise margin of the following circuits. The low noise margin allows other noise source to interfere with the circuit operation which creates problem in logic circuit operations. The transition region gain also approaches 1. To achieve sufficient gain for use in a digital circuit, the supply will be at least two times $\phi_T = kT/q$ (= 25 mV at room temperature). Thermal noise is a major issue for low-power operation. Below ϕ_T, the thermal noise is dominant and interferes with the device operation and potentially result in unreliable operation. This relation can be expressed as

$$V_{DDmin} \rangle 2 \ldots\ldots 4 \frac{kT}{q} \quad (8.1)$$

Equation 8.1 suggests that the only way to get CMOS inverters to operate below 100 mV is to reduce the ambient temperature – or in other words to cool the circuit. The cooling system also deserves some area which basically destroys the basic objective of scaling.

8.3 Ternary Logic Inverter

8.3.1 Standard Ternary Logic Inverter (STI)

In our previous chapter, we have presented the detailed analysis of standard ternary inverter (STI) based on QDGFET. The implementation of ternary inverters [6–10] requires three inverters: negative ternary inverter (NTI), standard ternary inverter (STI), and positive ternary inverter (PTI) where l_0, l_1, and l_2 should be the outputs respectively. The truth table for these three inverters is shown in Table 8.1.

The circuit diagram of a standard ternary inverter based on QDGFET is shown in Fig. 8.2 where QDGFETs are replaced by their sub-25-nm model. The transfer characteristic of the complement function in sub-nm range is shown in Fig. 8.3. Parameters for the transistors are shown in Table 8.2. The static current draw characteristic is shown in Fig. 8.4.

Table 8.3 shows the calculated rise/fall time of proposed inverter based on QDGFET compared to conventional architecture [11] which shows the improved performance of designed circuits (Fig. 8.5).

8.3.1.1 Noise Margin

There are four noise margins for ternary logic gates. One for logic level 0, two for logic level 1, and one for logic level 2 [11]. Different noise margins for a ternary logic gate can be defined as:

$$\begin{aligned} NM_2 &= V_{O2} - V_{I2} && \text{(a)} \\ NM_1{}^{+} &= V_{I1}{}^{+} - V_{O1}{}^{+} && \text{(b)} \\ NM_1{}^{-} &= V_{O1}{}^{-} - V_{I1}{}^{-} && \text{(c)} \\ NM_0 &= V_{I0} - V_{O0} && \text{(d)} \end{aligned} \quad (8.2)$$

Figure 8.6 shows different noise margins for a ternary logic gate. Table 8.4 shows different noise margins for STI based on QDGFET. Table 8.5 shows the power dissipation in standard ternary logic inverters based on QDGFETs compared to existing CMOS devices.

Table 8.1 Ternary logic inverter truth table

Input	STI	PTI	NTI
0	2	2	2
1	1	2	0
2	0	0	0

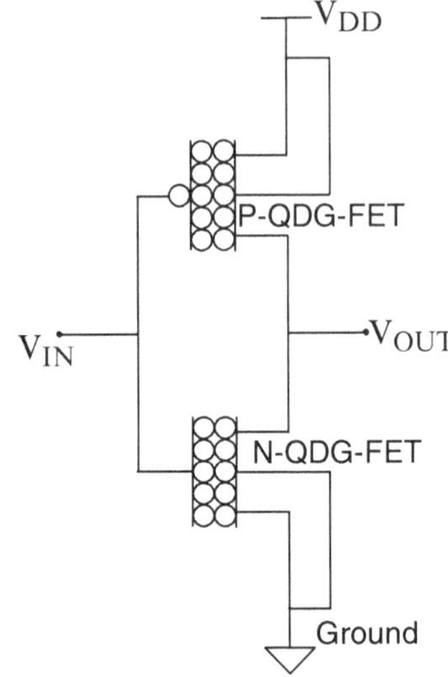

Fig. 8.2 CMOS architecture using quantum dot gate FET (QDGFET)

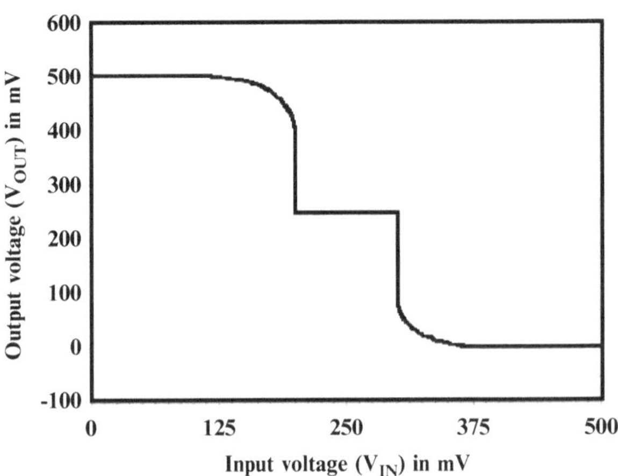

Fig. 8.3 Transfer characteristic of standard ternary inverter (STI) based on QDGFET

8.3.2 *Negative Ternary Logic Inverter (NTI) and Positive Ternary Logic Inverter (PTI)*

The circuit diagram for the negative ternary inverter (NTI) and the positive ternary inverter (PTI) is same as STI (Fig. 8.2) where the QDGFETs have different threshold voltages with 32 nm feature size parameters. Based on the threshold

Table 8.2 QDGFET parameters for sub-25-nm range

	N-QDGFET	P-QDGFET
Minimum L	32 nm	32 nm
Minimum W	32 nm	32 nm
V_T	100 mV	100 mV
V_{g1}	200 mV	200 mV
V_{g2}	300 mV	300 mV
V_{DD}	500 mV	500 mV
α	1.0	1.0

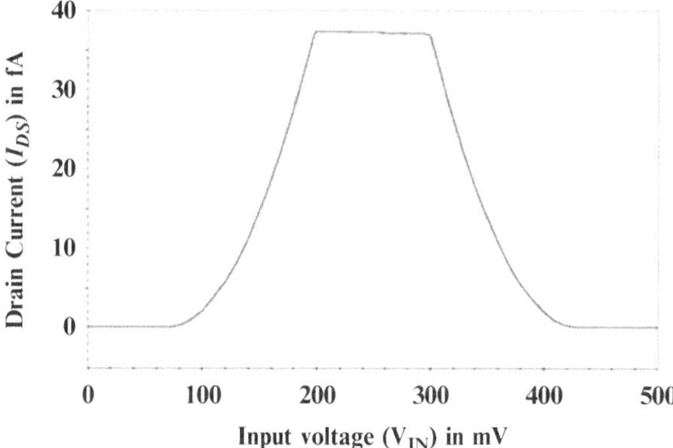

Fig. 8.4 Drain current (I_{DS}) versus input voltage (V_{IN}) transfer characteristics for QD complement function

Table 8.3 Rise/fall time for standard ternary logic inverter (STI)

Transition logic to logic		Rise/fall time in ns	
		Conventional	Proposed
1	2	150	42
2	1	75	38
1	0	150	28
0	1	80	34
2	0	190	112
0	2	190	125

voltage of PMOS and NMOS, the inverter can behave as a NTI or as a PTI. For NTI, the threshold voltage of PMOS is −0.3 V and that of NMOS is 0.11 V. When the input voltage is below 0.11 V, the NMOS is off and output is 0.5 V. When the input voltage is above 0.11 V, NMOS is on and the output is zero. For PTI implementation, the threshold voltage of NMOS and PMOS should be 0.3 and −0.2 V, respectively. Therefore, when the input voltage is more than 0.3 V, the lower transistor is on and output is zero. The transfer characteristics of NTI and PTI are shown in Fig. 8.7.

Table 8.6 shows the rise and fall time for NTI. Figure 8.8 shows the comparison of different delays of NTI based on conventional FETs and QDGFET. Table 8.7 shows the rise and fall time for NTI. Figure 8.9 shows the comparison of different delays of NTI based on conventional FETs and QDGFET. Table 8.8 shows the different noise margin for NTI and PTI. Table 8.9 shows the power dissipation for different inverters based on conventional FETs and QDGFETs.

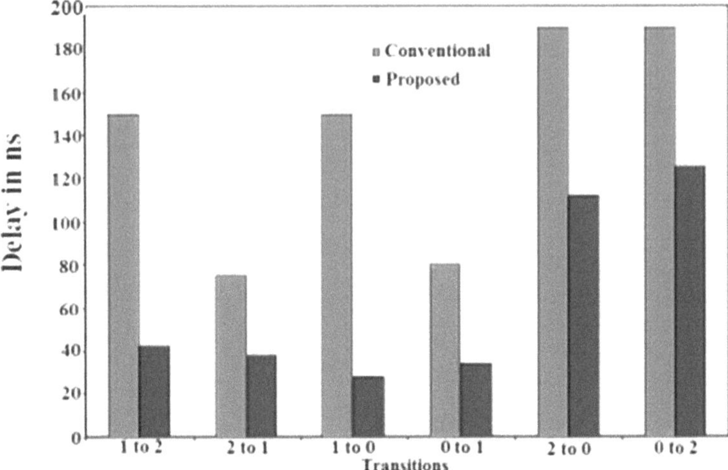

Fig. 8.5 Comparison of different delays of STI based on conventional FETs and QDGFETs

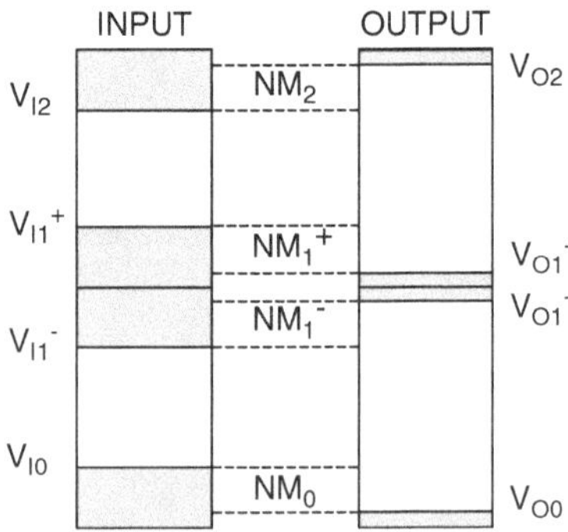

Fig. 8.6 Different noise margins for ternary logic

Table 8.4 Noise margin for STI

NM_0	55 mV
NM_{1-}	50 mV
NM_{1+}	50 mV
NM_2	55 mV

Table 8.5 Power dissipation

	Conventional	QDGFET based
STI	2.56 mili-watt	7.2 p-watt

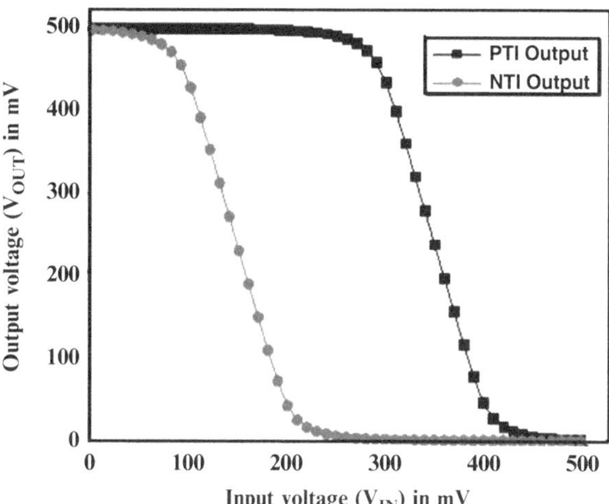

Fig. 8.7 Output voltage (V_{OUT}) versus input voltage (V_{IN}) transfer characteristics of PTI and NTI based on QDGFET

Table 8.6 Rise/fall time for NTI

Transition logic to logic		Rise/fall time in ns	
		Conventional	Proposed
0	2(t_{PLH})	360	162
2	0(t_{PHL})	35	10

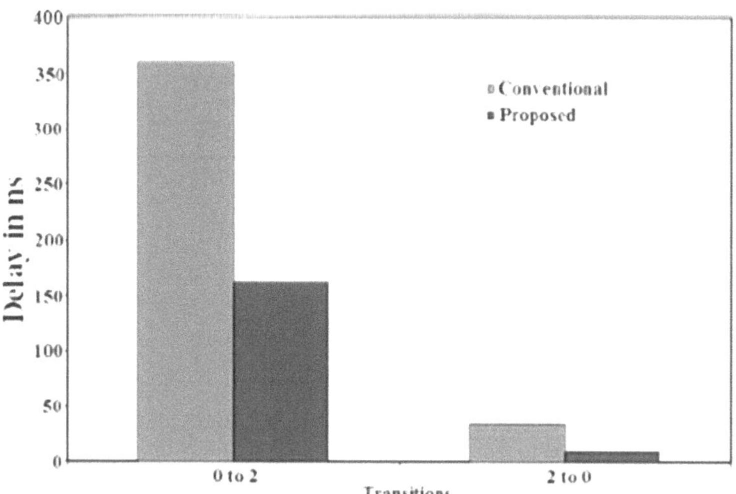

Fig. 8.8 Comparison of different delays of NTI based on conventional FETs and QDGFETs

Table 8.7 Rise/fall time for PTI

Transition logic to logic		Rise/fall time in ns Conventional	Proposed
0	2(t_{PLH})	35	15
2	0(t_{PHL})	360	110

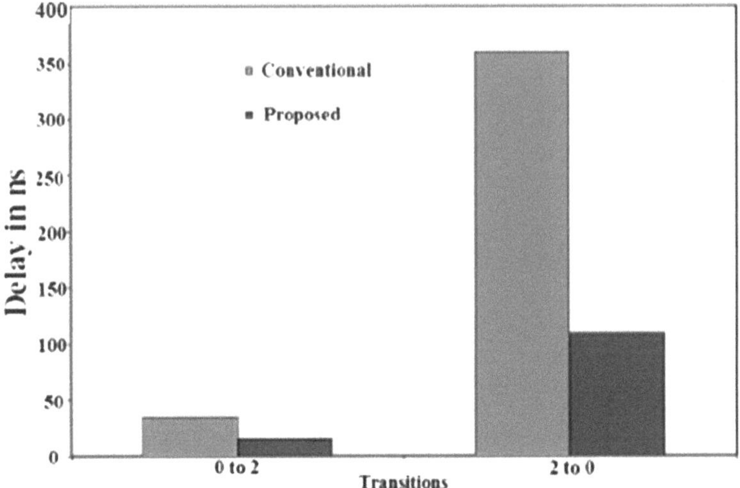

Fig. 8.9 Comparison of different delays of PTI based on conventional FETs and QDGFETs

Table 8.8 Noise margin for NTI and PTI

	NM_H in mV	NM_L in mV
PTI	52	243
NTI	254	40

Table 8.9 Power dissipation

	Conventional	QDGFET based
STI	2.56 mili-watt	7.2 p-watt
PTI	2.51 mili-watt	1.5 p-watt
NTI	2.51 mili-watt	6.3 p-watt

8.4 Two-Input Ternary Functions

The complementary function shows the use of QDGFETs in a three-state single input function. In order to build more useful circuits with three states or ternary logic, it is necessary to construct a set of two-input gate functions. With binary logic, the basic two-input functions are AND and OR. With ternary logic, we can build analogs of ANDs and ORs that have slightly different meanings. Kleene developed a set of functions for three-valued logic that has been termed Kleene's

Table 8.10 Kleene's logic

A	B	$A \wedge B$ (A AND B)	$A \vee B$ (A OR B)
F	F	F	F
F	U	F	U
F	T	F	T
U	F	F	U
U	U	U	U
U	T	U	T
T	F	F	T
T	U	U	T
T	T	T	T

logic [12]. This logic has three states: true (T), false (F), and undefined (U). Kleene also defined the functions $A \wedge B$ and $A \vee B$ as shown in Table 8.10. In the context of our state assignment, F corresponds to state 0, U corresponds to state 1, and T corresponds to state 2. In this case, undefined is a distinct third state or the intermediate state. We have constructed these functions with the above realization of the undefined states for both NAND and NOR functions as described below.

8.4.1 Ternary Logic NAND

8.4.1.1 Operation Principle

The NAND circuit diagram using a QDGFET is shown in Fig. 8.10. The circuit is similar to the normal CMOS NAND gate where normal FETs are replaced by quantum dot gate FETs. The circuit operation can be explained as follows. When either of the inputs is 0, the output is 2 because at least one of the P-QDGFETs is on, thus providing a path to V_{DD}, and at least one of the N-QDGFETs is off, thus cutting off a path to ground. When both of the inputs are 2, the output is 0 because both N-QDGFETs are on thus providing a path to ground. This behavior is identical to that of a conventional CMOS NAND gate.

On the other hand, for the undefined output cases, in Table 8.10 the behavior of the QDGFET NAND gate is more interesting. For this analysis, we assume that the "1" voltage corresponds to $\frac{V_{g2}-V_{g1}}{2}$. In other words, the "1" voltage is in the middle of the intermediate region. The QDGFET can be fabricated such that $\frac{V_{g2}-V_{g1}}{2} = \frac{V_{DD}}{2}$ as is the case with the QDGFET transistors that we are simulating.

When A = 1 and B = 1, all four QDGFETS are in the intermediate region. Both P-QDGFETs are operating in saturation, and it can also be determined that the B N-QDGFET is in the linear region and the A N-QDGFET is in saturation. Thus, assuming $\alpha = 1$,

$$I_{DSp} = \beta_p \frac{\left(V_{DD} - V_i - V_{Teff}\right)^2}{2} \tag{8.3}$$

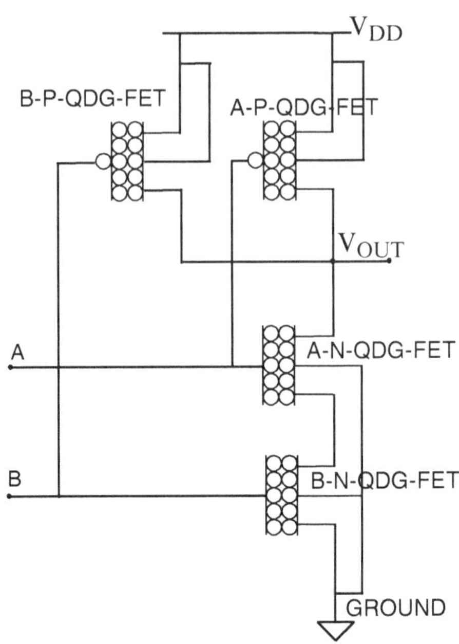

Fig. 8.10 Ternary logic NAND gate architecture

$$= \beta_p \frac{(V_{g1} - V_T)^2}{2} \tag{8.4}$$

$$I_{DSnA} = \beta_A \frac{(V_1 - V_{DB} - V_{Teff})^2}{2} \tag{8.5}$$

$$= \beta_A \frac{(V_{g1} - V_T)^2}{2} \tag{8.6}$$

$$I_{DSnB} = \beta_B \left(V_1 - V_{Teff} - \frac{V_{DB}}{2} \right) V_{DB} \tag{8.7}$$

$$= \beta_B \left(V_{g1} - V_T - \frac{V_{DB}}{2} \right) V_{DB} \tag{8.8}$$

where $V_1 = \frac{V_{g2} - V_{g1}}{2}, \beta_P = \frac{W_P}{L_P} C_o \mu_P, \beta_A = \frac{W_A}{L_A} C_o \mu_n, \beta_B = \frac{W_B}{L_B} C_o \mu_n$, and V_{DB} is the drain voltage of the B N-QDGFET.

Balancing the currents, we can set $2I_{DSp} = I_{DSnA} = I_{DSnB}$. Solving these equations, we arrive at the following:

$$V_{DB} = (V_{g1} - V_T) \left[1 - \sqrt{1 - \frac{\beta_A}{\beta_B}} \right] \tag{8.9}$$

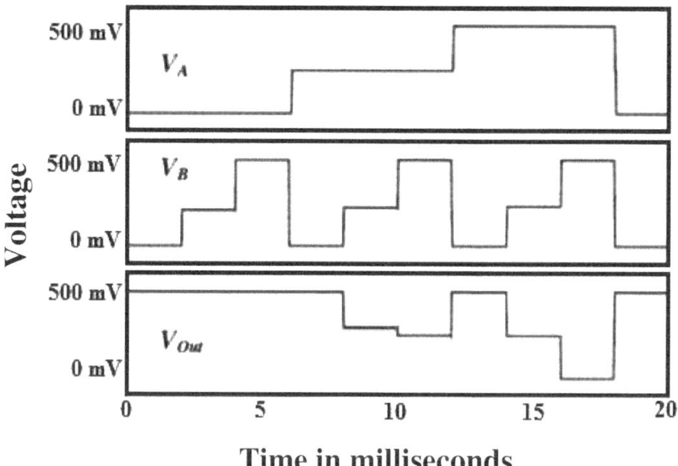

Fig. 8.11 Ternary logic NAND gate input-output waveforms

$$\beta_A = 2\beta_p \tag{8.10}$$

Because all four transistors are in the intermediate region, the current and output voltages are stable at this point. In order to have sufficient noise margin around this intermediate point, we need the V_1 +/−NM input to also provide a stable input. It would be expected that the intermediate region, V_{g1} to V_{g2}, would give the required noise margin, since V_1 is in the middle of that range. However, for transistor A, $V_{GS} = V_A - V_{DB}$. Therefore, to keep transistor A in the intermediate region, we need $V_A - V_{DB} > V_{g1}$. Thus, to preserve a large noise margin, we need V_{DB} to be as low as possible. Looking at the above equation, we see that $\frac{\beta_A}{\beta_B}$ ratio should be as low as possible. Using a low $\frac{\beta_A}{\beta_B}$ ratio also brings the output to V_{DD2} or V_1.

If the $\frac{\beta_A}{\beta_B}$ ratio is too low, such as when $\beta_A = \beta_B$, then V_{DB} becomes so high that $V_A - V_{DB} > V_{g1}$, and the A transistor is no longer in the intermediate region when $V_A = V_1$. In this case, the output voltage now becomes V_{DD} or logic value 2.

If we look at the other undefined outputs, namely, $V_A = 2, V_B = 1$, and $V_A = 1$, $V_B = 2$, we can see similar differences when we change the β ratios. When the β ratio is 1, the output voltage is V_1, and when the β ratio is low, the output voltage is 0. When the β ratio is in between 0.33 and 0.5, we get all combinations of output.

We have developed an AHDL (analog hardware description language) description of the QDGFET transistor model described before, and using Cadence's spectre simulator and this AHDL model, we have simulated the NAND gate with β ratio within 0.33 to 0.5.

The simulated output is shown in Fig. 8.11, and the truth table is shown in Table 8.11. The truth table in Table 8.11 is equivalent to a ternary NAND gate.

Table 8.11 Ternary NAND truth table

A	B	NAND OUTPUT
0	0	2
0	1	2
0	2	2
1	0	2
1	1	1
1	2	1
2	0	2
2	1	1
2	2	0

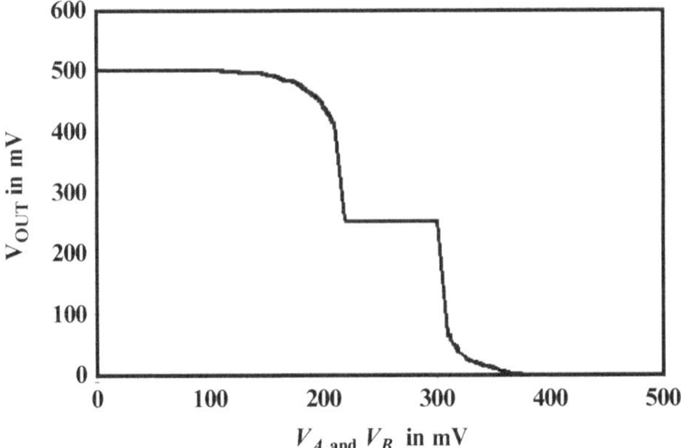

Fig. 8.12 Noise margin calculation for ternary logic NAND gate

8.4.1.2 Noise Analysis

Figure 8.12 shows the output voltage corresponds to the triangular waves in two inputs. From the plot, it can be determined that the intermediate state in the output is stable between 200 and 300 mV range of input voltage variation. Since the intermediate state is set to be 250 mV, this allows a noise margin of 50 mV.

8.4.2 *NAND as a Universal Logic Gate*

NAND gate can be used as a universal logic gate to implement other logic gates. We have demonstrated the implementation of other ternary logic gates using ternary logic NAND gate. The circuit is similar to conventional CMOS architecture. Figures 8.13 and 8.14 show the circuit diagram and the simulation results for implementation of different logic gates using ternary logic NAND gates.

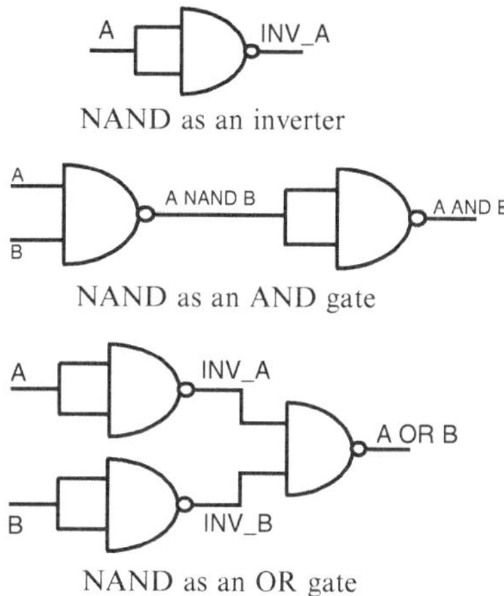

Fig. 8.13 Circuit diagram for different logic implementation using NAND

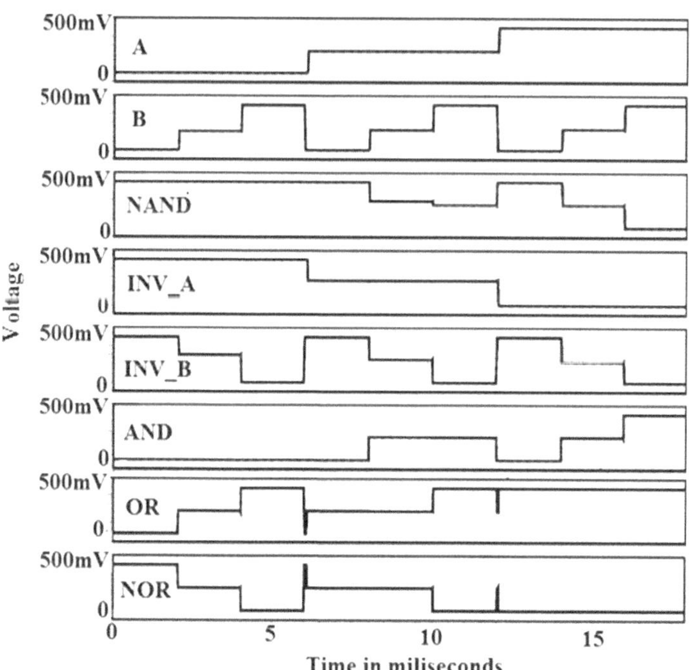

Fig. 8.14 Waveform of different ternary logic gates using ternary NAND gate

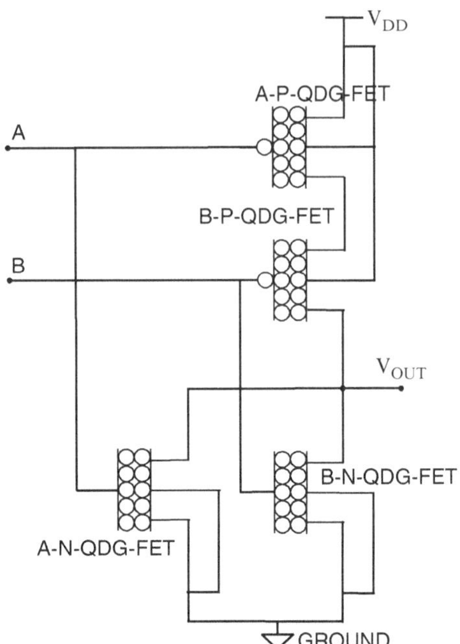

Fig. 8.15 Ternary NOR gate using QDGFET

8.4.3 *Ternary Logic NOR*

8.4.3.1 Operation Principle

The NOR circuit diagram using a QDGFET is shown in Fig. 8.15. The circuit is similar to the normal CMOS NOR gate where normal FETs are replaced by quantum dot gate FETs. The circuit operation can be explained as follows. When either of the inputs is 2, the output is 0 because at least one of the N-QDGFETs is on, thus providing a path to ground and at least one of the P-QDGFETs is off, thus cutting off a path to V_{DD}. When both of the inputs are 0, the output is 2 because both P-QDGFETs are on thus providing a path to V_{DD}. This behavior is identical to that of a conventional CMOS NOR gate.

As with the QDGFET NAND, the undefined output cases in Table 8.10 are more interesting. When A = 1 and B = 1, all four QDGFETS are in the intermediate region. Both N-QDGFETs are operating in saturation, and it can also be determined that the B P-QDGFET is in the linear region and the A P-QDGFET is in saturation. Thus, assuming $\alpha = 1$,

$$I_{DSn} = \beta_n \frac{\left(V_{g1} - V_T\right)^2}{2} \tag{8.11}$$

$$I_{DSpA} = \beta_{pA} \frac{\left(V_{g1} - V_T\right)^2}{2} \tag{8.12}$$

$$I_{DSpB} = \beta_{pB} \left(V_{g1} - V_T - \frac{\mathrm{V_{DD}} - V_{DB}}{2} \right) (\mathrm{V_{DD}} - V_{DB}) \tag{8.13}$$

where V_{DB} is the drain voltage of the B P-QDGFET.

Balancing the currents, we can set $2I_{DSn} = I_{DSpA} = I_{DSpB}$. Solving these equations, we arrive at the following:

$$V_{DB} = \mathrm{V_{DD}} - \left(V_{g1} - V_T\right) \left[1 - \sqrt{1 - \frac{\beta_{pA}}{\beta_{pB}}} \right] \tag{8.14}$$

$$\beta_{pA} = 2\beta_n \tag{8.15}$$

As with the QDGFET NAND, the operation is dependent on the $\frac{\beta_{pA}}{\beta_{pB}}$ ratio. When the ratio is low, V_{DB} is low, and the output is 1 when $V_A = V_B = V_1$ and the output is 2 when $V_A = 0/1$, $V_B = 1/0$. Likewise, when the ratio is 1, the output is 0 when $V_A = V_B = V_1$, and the output is 1 when $V_A = 0/1$, $V_B = 1/0$. For β ratio between 0.33 and 0.5, we get all combinations of outputs.

We also simulated the NOR gate, and the results are shown in Fig. 8.16 and the truth table is shown in Table 8.12. As with the QDGFET NAND gate, the truth table in Table 8.12 is equivalent to a ternary NOR gate.

8.4.3.2 Noise Analysis

Following the similar procedure like NAND gate, it can be determined that the intermediate state in the output of NOR gate is stable between 200 and 300 mV range of input voltage variation. Since the intermediate state is set to be 250 mV, this allows a noise margin of 50 mV.

8.4.4 NOR as a Universal Logic Gate

NOR gate can be used as a universal logic gate to implement other logic gates. We have demonstrated the implementation of other ternary logic gates using ternary logic NOR gate. The circuit is similar to conventional CMOS architecture. Figures 8.17 and 8.18 show the circuit diagram and the simulation results for implementation of different logic gates using ternary logic NOR gates.

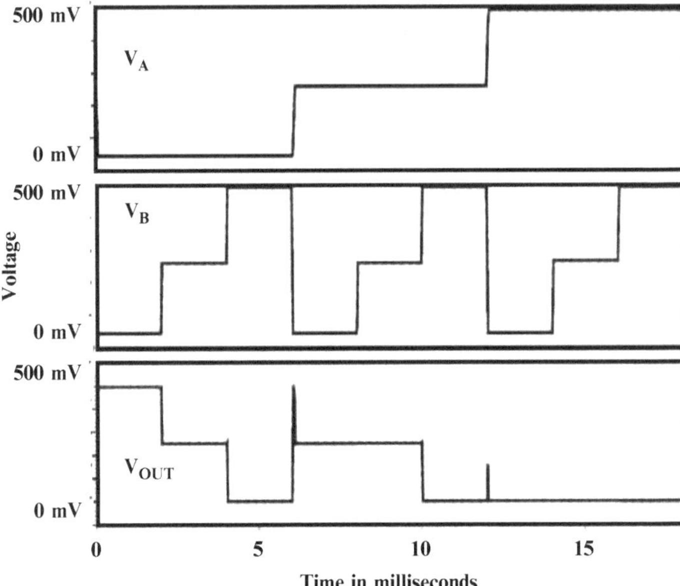

Fig. 8.16 Ternary logic NOR gate input-output waveforms

Table 8.12 Ternary NOR truth table

A	B	NOR output
0	0	2
0	1	1
0	2	0
1	0	1
1	1	1
1	2	0
2	0	0
2	1	0
2	2	0

8.4.5 Ternary Logic XOR Gate

A XOR function can be implemented by combining inverter, AND and OR gates. The simulated input-output waveform of the ternary XOR gate is shown in Fig. 8.19. The truth table for a ternary logic XOR gate is shown in Table 8.13. Different time delays are shown in Table 8.14.

8.4.5.1 Voltage

Table 8.15 shows the power dissipation of different ternary logic circuits based on QDGFET which are discussed in this paper.

Fig. 8.17 Circuit diagram for different logic implementation using NOR

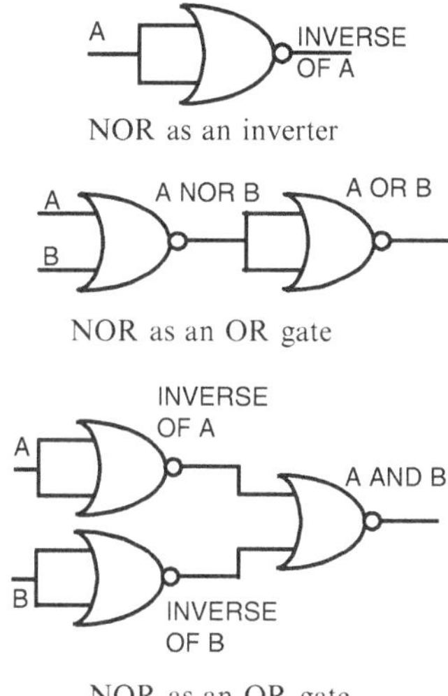

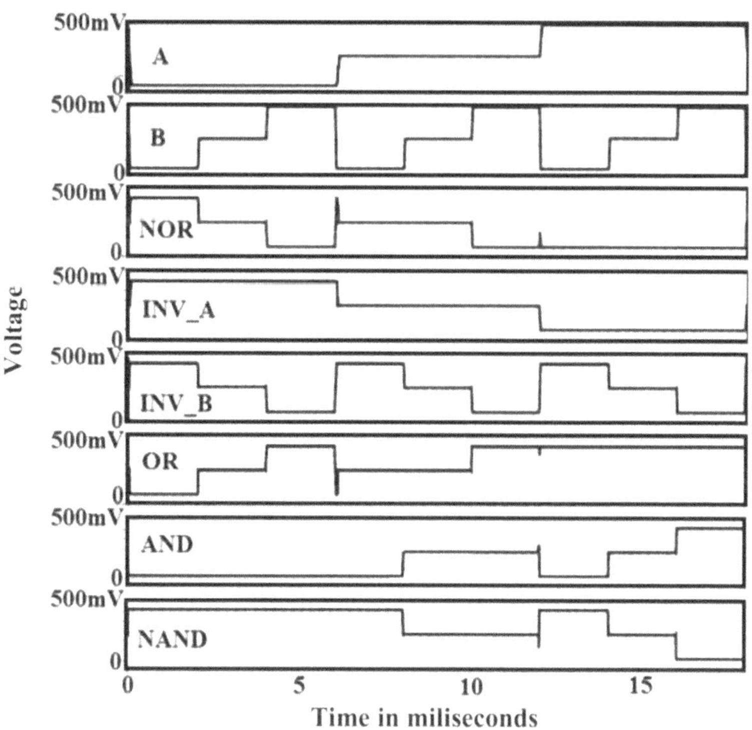

Fig. 8.18 Waveform of different ternary logic gates using ternary NOR gate

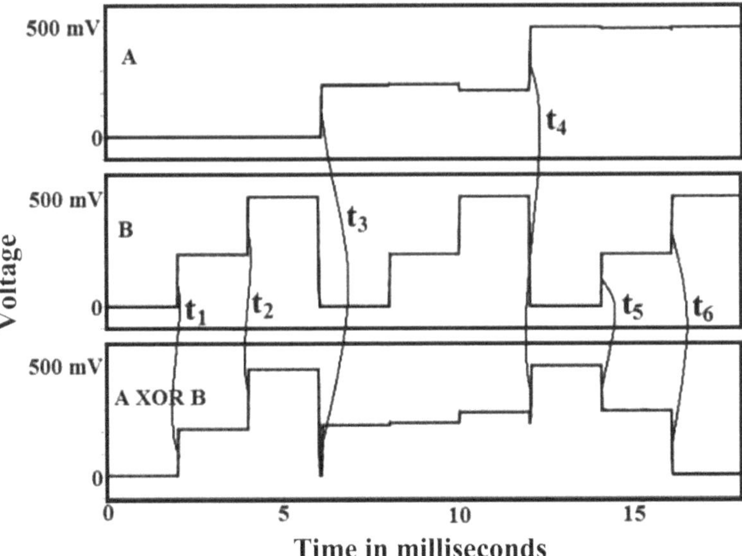

Fig. 8.19 Ternary logic XOR gate input-output waveforms

Table 8.13 Truth table for ternary logic XOR

A/B	0	1	2
0	0	1	2
1	1	1	1
2	2	1	0

Table 8.14 Different time delays in ternary logic XOR

Different delays	Time in picoseconds
t_1	0.3
t_2	0.8
t_3	0.2
t_4	1.1
t_5	0.1
t_6	0.5

Table 8.15 Power dissipation in different logic gates

Logic circuit	Power dissipation
Ternary AND	0.1 n-watt
Ternary OR	0.1 n-watt
Ternary XOR	0.9 n-watt

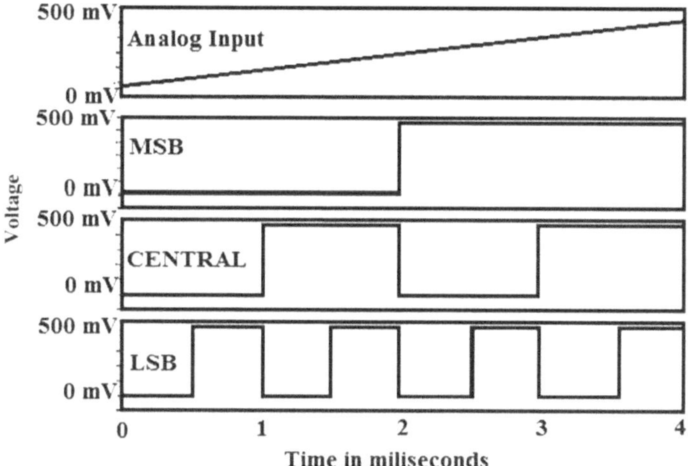

Fig. 8.20 Input-output waveforms digital-to-analog converter (DAC)

Table 8.16 Different performance parameters of the designed ADC in sub-25-nm range

Signal-to-noise ratio (SNR)	18.2348 dB
Signal-to-noise and distortion ratio (SNDR)	12.7508 dB
Total harmonic distortion (THD)	−14.1947 dB
Spurious frequency dynamic range (SFDR)	16.0453 dBFS
Error in number of bit (ENOB)	1.82571 bits

8.5 Three-Bit Analog-to-Digital Converter (ADC)

A flash ADC is the simplest and fastest ADC architecture. Detailed architecture of QDGFET-based flash ADC is already discussed in Chap. 7. In this section, three-bit flash ADC is designed following the same architecture with 25-nm QDGFET model.

8.6 Three-Bit Digital-to-Analog Converter (DAC)

The flash DAC based on 25-nm QDGFET model is designed in the same way following the same architecture discussed in Chap. 7. The input-output waveforms are shown in Fig. 8.20. The performance analysis is shown in Table 8.16.

The three-bit ADC output is shown in Fig. 8.21.

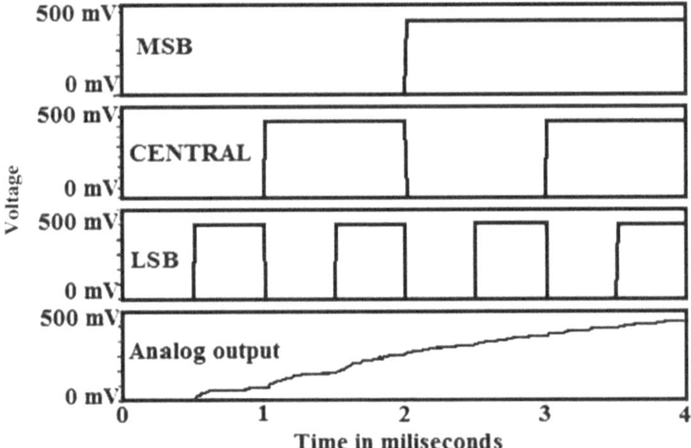

Fig. 8.21 Input-output waveform of three-bit ADC (25-nm QDGFET model)

Table 8.17 Truth table for decoder (First kind)

Input	J_2	J_1	J_0
0	0	0	1
1	0	1	0
2	1	0	0

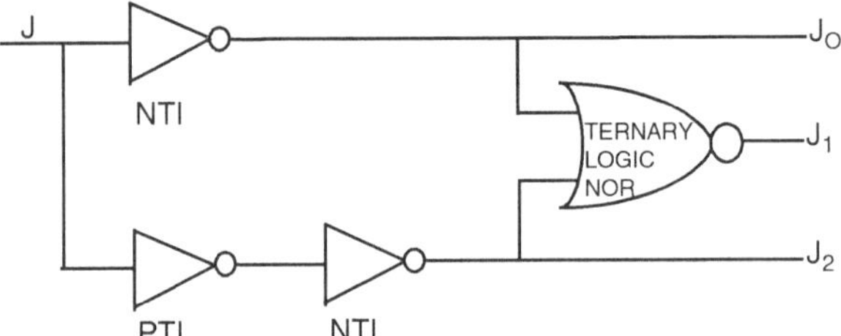

Fig. 8.22 Ternary logic decoder block diagram

8.7 Ternary Logic Decoder

8.7.1 *First Kind*

A ternary decoder is a one input three output combinational logic circuit. The output of a ternary decoder can be represented as

$$J_k(X) = \begin{cases} 2 & if \quad X = k \\ 0 & if \quad X \neq k \end{cases} \tag{8.16}$$

where $k = 0, 1, 2$ which corresponds to higher level (2), middle level (1), and lower level (0). The truth table for ternary logic decoder is shown in Table 8.17. The design of ternary decoder based on QDGFET-based circuits is shown in Fig. 8.22

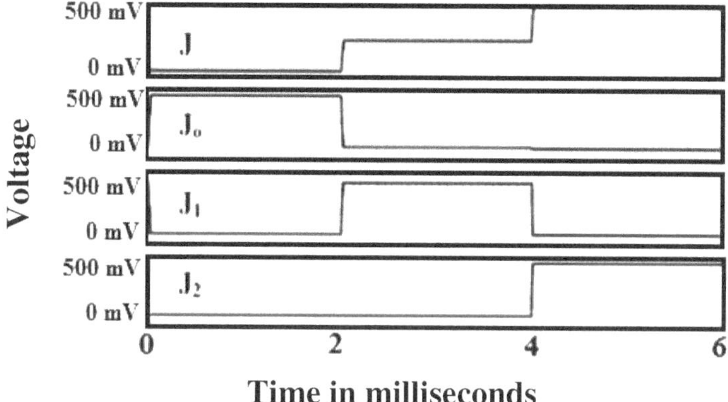

Fig. 8.23 Input-output waveform of designed decoder (First kind)

Table 8.18 Truth table for decoder (Second kind)

Input	J_1	J_0
0	0	0
1	0	1
2	1	0

Fig. 8.24 Ternary logic decoder block diagram (Second kind)

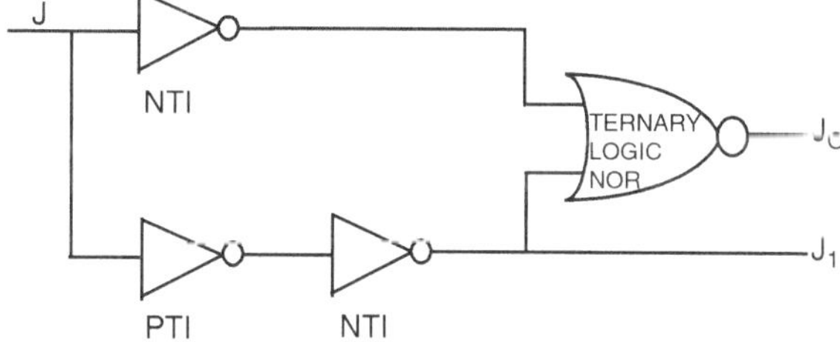

where two NTI gates, one PTI gate, and one ternary NOR gate are used. The output of the decoder is shown in Fig. 8.23.

8.7.2 Second Kind

The truth table for the second kind of decoder is shown in Table 8.18. The circuit diagram of the decoder is shown in Fig. 8.24. The input-output waveform of the designed decoder is shown in Fig. 8.25.

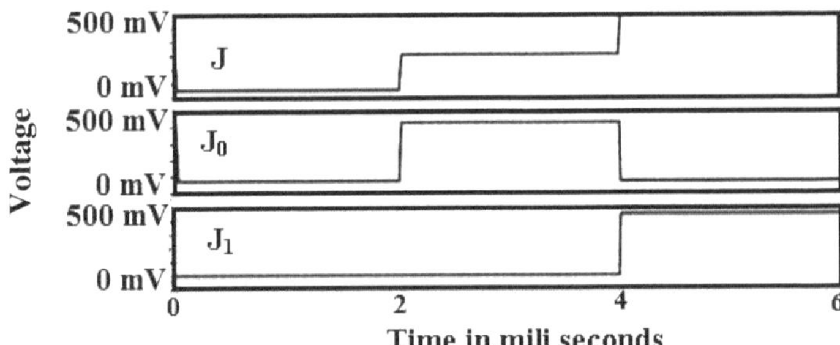

Fig. 8.25 Input-output waveform of designed decoder

References

1. Yang, F.-L.: 25 nm CMOS Omega FETs. In: IEDM Technical Digest, pp. 255–258, Dec 2002
2. Chang, Y.-C., et al.: A 25-nm gate-length FinFET transistor module for 32nm node. In: IEDM Technical Digest, pp. 12.2.1–12.2.4, Dec 2009
3. Rim, K.: Scaling of strain-induced mobility enhancements in advanced CMOS technology. In: ICSICT 2008, pp. 105–108, Oct 2008
4. Kuhn, K.: Moore's law past 32nm: future challenges in device scaling. In: 13th International Workshop on Computational Electronics, IWCE '09, pp. 1–6, 2009
5. Chandrakasan, A.P., et al.: Low-power CMOS digital design. IEEE J. Solid State Circuits **27**(4), 473–483 (1992)
6. Liu, D., Svensson, C.: Power consumption in CMOS VLSI chips. IEEE J. Solid State Circuits **29**(6), 663–670 (1994)
7. Wojcik, A.S., Fang, K.: On the design of three-valued asynchronous modules. IEEE Trans. I Comput. **C-29**(10), 889–898 (1980)
8. Wu, C., Huang, H.: Design and application of pipelined dynamic CMOS ternary logic and simple ternary differential logic. IEEE J. Solid State Circuits **28**(8), 895–906 (1993)
9. Raychowdhury, A., Roy, K.: A novel multiple-valued logic design using ballistic carbon nanotube FETs. In: Proceedings of International Symposium on Multiple-Valued Logic. Toronto, Canada, pp. 14–19, May 2004
10. Baba, T., Uemura, T.: Development of InGaAs-based multiple-junction surface tunnel transistors for multiple-valued logic circuits. In: Proceedings of International Symposium on Multiple-Valued Logic. Fukuoka, Japan, pp. 7–12 (1998)
11. Balla, P.C., Antoniou, A.: Low power dissipation MOS ternary logic family. IEEE J. Solid State Circuits **19**(5), 739–749 (1984)
12. Kleene, S.C.: Introduction to metamathematics, pp. 332–340. North-Holland, Amsterdam (1952)

Chapter 9
Conclusions

This chapter summarizes the conclusions of this book.

9.1 Conclusions

This book discusses the fabrication, modeling, and application of quantum dot gate field-effect transistors (QDGFETs). The intermediate state is observed for different combinations of quantum dots and tunnel insulators, as well as different types of substrates.

As the feature size decreases, the gate dielectric thickness needs to decrease in order to raise the gate capacitance and thereby the drive currents and the device performance. When the dielectric thickness is below 2 nm, the leakage current due to tunneling of charge carriers increases drastically. Development efforts have focused on finding a material with a requisitely high dielectric constant that allows increased gate capacitance without the concomitant leakage effects and can be easily integrated into the existing silicon manufacturing process. This book introduces a new kind of lattice-matched high-κ dielectric materials as gate insulator.

The data presented in this paper shows the observation of an intermediate state "i" in a quantum dot gate silicon FET using a lattice-matched ZnS-ZnMgS gate insulator layer. This presented device simulation model using self-consistent Poisson, Schrödinger, and tunnel rate solvers describes the generation of intermediate state between the two stable states (on and off).

The presence of the intermediate state in the transfer characteristics of a QDGFET fabricated on silicon-on-insulator substrate is also observed. The floating body of the SOI wafer reduces the punch-through effect of the field-effect transistor. The reduced leakage current between the source and the drain region decreases the off current of the device and increases the on-to-off ratio of the device. The use of SOI wafer also improves the subthreshold slope of the device because of less leakage current. The distinct intermediate state and improved device characteristic

S. Karmakar, *Novel Three-state Quantum Dot Gate Field Effect Transistor: Fabrication, Modeling and Applications*, DOI 10.1007/978-81-322-1635-3_9,

of the QDGFET on SOI substrates improves the device speed as well as the noise margin.

The distinct state and the improved performance of the QDGFET on SOI wafer will help to design different logic circuits based on this FET. The improved noise margin will solve many logic level problems in ternary logic circuits. The improved subthreshold slope will increase the speed of the logic circuits based on this device. The improved noise margin will also give a huge thrust in multivalued logic circuit design.

This book also manifested the successful fabrication of an NMOS inverter circuit based on the quantum dot gate field-effect transistor. The compatibility between the fabricated device characteristics and the circuit model will help us to understand and design different complex logic circuits based on QDGFET and QDNMOS inverters. The generation of an additional state in the output of QDNMOS inverter will make it an important circuit element in multivalued logic circuit and help us to fabricate different complex circuits based on conventional NMOS technology.

Successful circuit models for quantum dot gate FET (QDGFET) and quantum dot gate nonvolatile memory (QDNVM) based on BSIM are introduced in this thesis. Different fundamental multivalued logic building blocks such as ternary logic inverter, NAND, NOR, and XOR gates are implemented. Ternary logic to binary logic conversion using ternary logic decoder is also implemented. Ratioless property and static current problem of QDGFET-based circuits are also manifested. Analog circuits such as three-bit and six-bit analog-to-digital converters (ADCs) and digital-to-analog converters (DAC) are also designed.

The controllable threshold voltage of QDNVMs makes them effective for ADC circuit design applications. Precise control of the comparator reference voltage by controlling the threshold voltage of quantum dot gate nonvolatile memory makes this circuit free from the R-2R ladder problem. In addition, the numbers of transistors per comparator decrease from 32 to 1 with respect to conventional circuit design architecture. In addition to other advantages, the use of less number of circuit elements to implement complicated circuit based on QDNVM can make them promising circuit elements in the future.

About the Author

Dr. Karmakar completed his PhD in Electrical Engineering from University of Connecticut in the year 2011. His thesis was based on "Novel Three State Quantum Dot Gate Field Effect Transistor: Fabrication, Modeling and Applications". Currently he is working as an Engineer in Intel Corporation, Hillsboro, Oregon, USA. He has seven years experience in semiconductor device fabrication and circuit modelling. He has modified different photolithography processes and metal organic chemical vapor deposition (MOCVD) techniques to fabricate various types of semiconductor devices like quantum dot gate field effect transistor (QDGFET), quantum dot gate nonvolatile memory (QDNVM), quantum dot channel field effect transistors (QDCFET), Solar Cells etc. for different projects. He developed successful circuit model for the above mentioned devices and designed different complicated circuit based on those circuit models for multi-valued logic implementation. He has expertise in designing different types of masks for photolithography as well as E-Beam lithography. Dr. Karmakar has also published more than 30 papers in international peer-reviewed journals and Conference proceedings.

S. Karmakar, *Novel Three-state Quantum Dot Gate Field Effect Transistor: Fabrication, Modeling and Applications*, DOI 10.1007/978-81-322-1635-3,

Index

S. Karmakar, *Novel Three-state Quantum Dot Gate Field Effect Transistor: Fabrication, Modeling and Applications*, DOI 10.1007/978-81-322-1635-3,

Zeitfracht Medien GmbH
Ferdinand-Jühlke-Straße 7
99095 Erfurt, Deutschland
produktsicherheit@kolibri360.de